Springer Theses

Recognizing Outstanding Ph.D. Research

Aims and Scope

The series "Springer Theses" brings together a selection of the very best Ph.D. theses from around the world and across the physical sciences. Nominated and endorsed by two recognized specialists, each published volume has been selected for its scientific excellence and the high impact of its contents for the pertinent field of research. For greater accessibility to non-specialists, the published versions include an extended introduction, as well as a foreword by the student's supervisor explaining the special relevance of the work for the field. As a whole, the series will provide a valuable resource both for newcomers to the research fields described, and for other scientists seeking detailed background information on special questions. Finally, it provides an accredited documentation of the valuable contributions made by today's younger generation of scientists.

Theses may be nominated for publication in this series by heads of department at internationally leading universities or institutes and should fulfill all of the following criteria

- They must be written in good English.
- The topic should fall within the confines of Chemistry, Physics, Earth Sciences, Engineering, and related interdisciplinary fields such as Materials, Nanoscience, Chemical Engineering, Complex Systems, and Biophysics.
- The work reported in the thesis must represent a significant scientific advance.
- If the thesis includes previously published material, permission to reproduce this must be gained from the respective copyright holder (a maximum 30% of the thesis should be a verbatim reproduction from the author's previous publications).
- They must have been examined and passed during the 12 months prior to nomination.
- Each thesis should include a foreword by the supervisor outlining the significance of its content.
- The theses should have a clearly defined structure including an introduction accessible to new PhD students and scientists not expert in the relevant field.

Indexed by zbMATH.

Hiroki Arisawa

A Study on Spin Mechanics Phenomena in Spin-Lattice Coupled Systems

Doctoral Thesis accepted by
Tohoku University, Sendai, Japan

Springer

Author
Dr. Hiroki Arisawa
Department of Applied Physics
University of Tokyo
Tokyo, Japan

Supervisor
Prof. Eiji Saitoh
Department of Applied Physics
University of Tokyo
Tokyo, Japan

ISSN 2190-5053 ISSN 2190-5061 (electronic)
Springer Theses
ISBN 978-981-97-8641-1 ISBN 978-981-97-8642-8 (eBook)
https://doi.org/10.1007/978-981-97-8642-8

This Springer imprint is published by the registered company Springer Nature Singapore Pte Ltd.
The registered company address is: 152 Beach Road, #21-01/04 Gateway East, Singapore 189721, Singapore

Supervisor's Foreword

The interaction between magnetization and mechanical strain in magnets has been a central issue in magnetism for a long time. A notable example is the magneto-volume effect in a ferromagnet. Since the discovery of Invar alloys, where magneto-volume changes compensate for thermal expansion, extensive studies on the magneto-volume effect have led to significant progress in the physics of spin fluctuation and electronic correlation.

This book highlights the research on spin-mechanical interactions between magnetization and strain, conducted by Hiroki Arisawa. Using spintronics knowledge, he demonstrated that the magneto-volume effect can be triggered by injecting a spin current into a magnet. He also investigated the interplay between magnetization and mechanical vibrations in a magnetic mechanical oscillator. The work presented in this book sheds light on the fundamental physics of magnetoelasticity from the perspective of spintronics.

Tokyo, Japan
June 2024

Prof. Eiji Saitoh

Parts of this thesis have been published in the following journal articles:

1. H. Arisawa, S. Daimon, Y. Oikawa, Y.-J. Seo, K. Harii, K. Oyanagi, and E. Saitoh, Magnetomechanical sensing based on delta-E effect in $Y_3Fe_5O_{12}$ micro bridge, Appl. Phys. Lett. **114**, 122402, (2019).
2. H. Arisawa, H. Shim, S. Daimon, T. Kikkawa, Y. Oikawa, S. Takahashi, T. Ono, and E. Saitoh, Observation of spin-current striction in a magnet, Nat. Commun. **13**, 2440 (2022).

Acknowledgements

I would like to thank Prof. Eiji Saitoh for giving me the opportunity to join his group and for special support in carrying out my research. I am also grateful to Dr. Shunsuke Daimon, Dr. Takashi Kikkawa, Dr. Yasuyuki Oikawa, Dr. Saburo Takahashi, and Dr. Seo Yong-Jun for fruitful discussions and support for my research. I would like to express my gratitude to all former and present members of Saitoh laboratory for giving me a lot of encouragement.

I would like to thank the collaborators Prof. Takahito Ono and Dr. Hang Shim for supporting my experiments and valuable discussions.

I would like to thank Prof. Kozo Fujiwara for supporting my research activities and fruitful discussions.

I would like to thank my Ph.D. thesis committee members, Prof. Eiji Saitoh, Prof. Kozo Fujiwara, Prof. Masaki Fujita, Prof. Yoshiro Hirayama, and Assoc. Prof. Joji Nasu, for kindly accepting to review my thesis and valuable discussions.

Finally, I would like to express my gratitude to my parents and brother who have supported me a lot.

Contents

Chapter 1
Research Background

1.1 Introduction and Research Motivation

Spintronics is a research field that aims to investigate physical properties and their applications by making use of spin angular momentum of electrons in solids. Over the years, interconversion phenomena between spin angular momentum, charge, and heat have been explored extensively [1–4]. Spin Hall effect [1] and spin Seebeck effect [3] are typical examples, which electrically and thermally generate a spin current, a flow of spin angular momentum, respectively.

In recent years, the field of spin mechanics [5, 6] has attracted much attention as a research area on spin interconversion phenomena, which explores the cross-correlation between spin angular momentum and mechanical motion in solids. Spin angular momentum, lattice motion, and mechanical motion interact with each other via magnetoelastic coupling, and a spin-mechanical coupling is realized [7, 8].

The purpose of the present research is to investigate spin-mechanical coupled phenomena and to clarify their physical mechanism. For this aim, in this study, we performed two experiments in the following. In the first research, we tried to detect the coupling between the magnetization and the mechanical vibration in a micro magnetic mechanical oscillator. A micro mechanical vibration device such as a cantilever enables highly sensitive mechanical detection due to its high Q factor [9]. In addition, the progress of microfabrication technologies in recent years has made it possible to fabricate a micro mechanical device from magnetic materials [10]. Therefore, by imposing magnetism to the micro mechanical device, we can directly detect the spin-mechanical coupling in the device with high sensitivity. In the second research, we challenged to demonstrate the volume manipulation effect by using a spin current in a spin-lattice coupled system. Alloy systems of rare earth elements and iron are known to exhibit large magnetoelastic coupling and giant magnetostriction [11]. By using the spin-lattice coupled system, we explored spin-current-induced spin mechanical phenomena.

H. Arisawa, *A Study on Spin Mechanics Phenomena in Spin-Lattice Coupled Systems*, Springer Theses, https://doi.org/10.1007/978-981-97-8642-8_1

1.2 Structure of Thesis

The structure of the thesis is shown below. In the previous section, in this chapter, we introduced the research background of the field of spin mechanics and described the main aim of the present research. After this section, we explain the basic physics on spin systems, lattice systems, and spin-lattice coupled systems.

In Chap. 2, we explain the physical properties and the fabrication methods of the samples used in the experiments. We also show the measurement setup of the spin-lattice coupled phenomena, which is realized by using a laser Doppler vibrometer.

In Chap. 3, we discuss experimental results on the cross-correlation effect between the magnetization and the mechanical vibration in a micro magnetic mechanical oscillator [12]. By fabricating a micro mechanical oscillator made from a magnetic insulator $Y_3Fe_5O_{12}$ (YIG) and measuring the mechanical response of the YIG oscillator, we found that the resonant mechanical vibration is modulated in accordance with the magnetization process in the oscillator. Systematic experiments and theoretical analysis revealed that the observed changes in the mechanical properties in the oscillator originate from the Young's modulus modulation due to the magnetization rotation.

In Chap. 4, we discuss a volume modulation effect in a magnet induced by a spin current [13]. By injecting a spin current into a giant magnetostrictive material $Tb_{0.3}Dy_{0.7}Fe_2$, we demonstrated that a $Tb_{0.3}Dy_{0.7}Fe_2$ film exhibits the thickness change in response to the spin current injection. By comparing the experimental results and the magnetoelastic theoretical model, we found that the observed thickness change of the $Tb_{0.3}Dy_{0.7}Fe_2$ film is attributed to the magnetization fluctuation change in the film induced by the spin current injection.

In Chap. 5, we summarize the obtained experimental results and conclusions.

1.3 Spin Dynamics

One of the basic concepts in spintronics is a flow of spin angular momentum, a spin current. Since the discovery of the spin current generation using an electric current and its inverse effect, various spintronic phenomena based on a spin current have been investigated. In this section, we define a spin current and explain the spin dynamics associated with it.

1.3.1 Conduction-Electron Spin Current

A conduction-electron spin current is a flow of spin angular momentum carried by conduction electrons in a metal. We define a conduction-electron spin current $\hat{j}_s$ by the continuity equation of the spin angular momentum density $\mathbf{S}(\mathbf{r})$:

$$\frac{\partial \mathbf{S}(\mathbf{r})}{\partial t} = -\nabla \cdot \hat{j}_s, \tag{1.1}$$

where $\hat{j}_s$ is a second-order tensor about the direction of a spin current and the direction of the spin angular momentum carried by the spin current. By using the Heisenberg equation and the Hamiltonian $\mathcal{H}$ of free electrons

$$\frac{\partial \mathbf{S}(\mathbf{r})}{\partial t} = i\hbar[\mathbf{S}(\mathbf{r}), \mathcal{H}] \tag{1.2}$$

$$\mathcal{H} = \int d\mathbf{r} \frac{\hbar^2}{2m} \sum_{\sigma=\uparrow,\downarrow} \nabla c_\sigma^\dagger(\mathbf{r}) \nabla c_\sigma(\mathbf{r}), \tag{1.3}$$

we can obtain the explicit form of $\hat{j}_s$:

$$\hat{j}_s = \frac{i\hbar^2}{4m} \sum_{\sigma,\sigma'=\uparrow,\downarrow} \{\nabla(c_\sigma^\dagger)(\hat{\sigma})_{\sigma\sigma'} c_{\sigma'} - c_\sigma^\dagger(\hat{\sigma})_{\sigma\sigma'} \nabla(c_{\sigma'})\} \tag{1.4}$$

Here, $\hbar$, m, c_σ, σ, and $\hat{\sigma}$ are Dirac constant, the mass of an electron, the annihilation operator of an electron, the spin angular momentum, and the Pauli matrix, respectively. When we take the quantization axis in the z direction, we can derive the z component of the spin current tensor $\hat{j}_s$ as

$$\mathbf{j}_s = \frac{\hbar}{2}(\mathbf{j}_\uparrow - \mathbf{j}_\downarrow), \tag{1.5}$$

where

$$\mathbf{j}_\sigma = \frac{i\hbar}{2m}(\nabla(c_\sigma^\dagger)c_\sigma - c_\sigma^\dagger \nabla(c_\sigma)) \tag{1.6}$$

is a flow of electrons with the spin angular momentum $\sigma = \uparrow, \downarrow$. Equation (1.5) means that the conduction-electron spin current can be interpreted as the difference between the up-spin electron flow $\mathbf{j}_\uparrow$ and the down-spin electron flow $\mathbf{j}_\downarrow$ (Fig. 1.1).

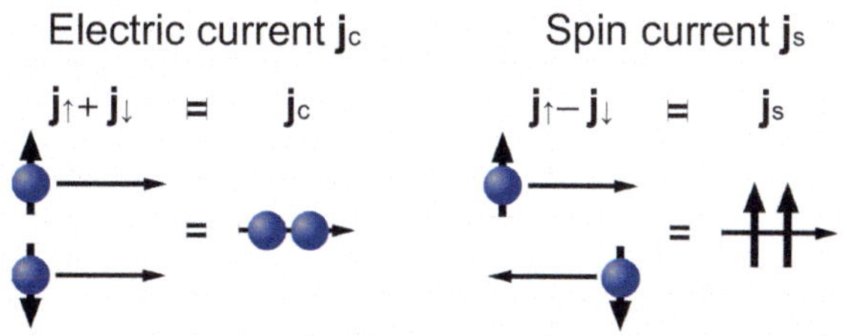

Fig. 1.1 A schematic illustration of an electric current $\mathbf{j}_c$ and a conduction-electron spin current $\mathbf{j}_s$. $\mathbf{j}_s$ is defined as the difference between an up spin current $\mathbf{j}_\uparrow$ and a down spin current $\mathbf{j}_\downarrow$

1.3.2 Spin Hall Effect

The spin Hall effect is a phenomenon where an electric current is converted into a spin current (Fig. 1.2), which satisfies the following equation:

$$\mathbf{j}_s \propto \theta_{SH}\boldsymbol{\sigma} \times \mathbf{j}_c \tag{1.7}$$

Here, $\mathbf{j}_s$, $\mathbf{j}_c$, and $\boldsymbol{\sigma}$ are the spin current density, the charge current density, and spin angular momentum, respectively. θ_{SH} is the spin Hall angle, which determines the sign and the strength of the spin current due to the spin Hall effect [14]. The value of the spin Hall angle depends on the strength of the spin-orbit interaction in each material. Pt (Cu) exhibits a large (small) spin Hall angle, reflecting that the spin-orbit interaction is proportional to the atomic number. W is known to exhibit a spin Hall angle with an opposite sign compared to Pt. In the case of Pt ($\theta_{SH} > 0$), as shown in Fig. 1.2, an electric current in the $+x$ direction is converted into a spin current in the $-y$ direction via the spin Hall effect, whose spin is polarized in the $+z$ direction.

1.3.3 LLG Equation

Magnetization dynamics in a ferromagnet is described by the LLG equation [15]:

$$\frac{d\mathbf{M}}{dt} = -\gamma\mathbf{M} \times \mathbf{H}_{\text{eff}} + \frac{\alpha}{M_s}\mathbf{M} \times \frac{d\mathbf{M}}{dt}, \tag{1.8}$$

where $\mathbf{M}$, $\mathbf{H}_{\text{eff}}$, α, and M_s are the magnetization in a magnet, the effective magnetization, the Gilbert damping constant, and the saturation magnetization of the magnet, respectively. The first term on the right hand side corresponds to the magnetic torque

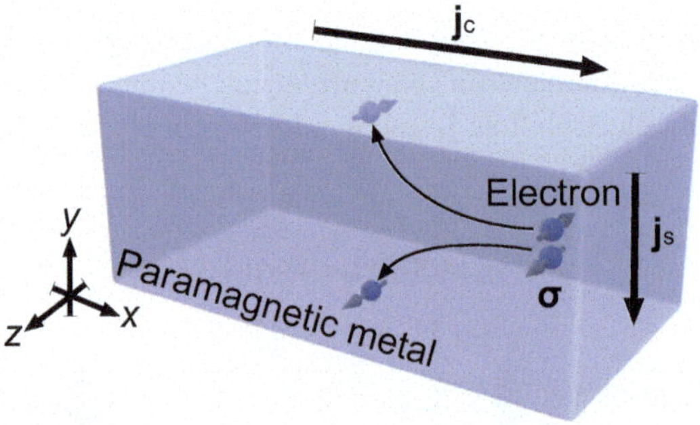

Fig. 1.2 A schematic illustration of the spin Hall effect. When an electron with the spin angular momentum $\boldsymbol{\sigma}$ along the $+z$ ($-z$) direction flows in the $-x$ direction, the direction of the electron flow is changed to the $-y$ ($+y$) direction via the spin orbit coupling, and an electric current $\mathbf{j}_c$ is converted into a spin current $\mathbf{j}_s$

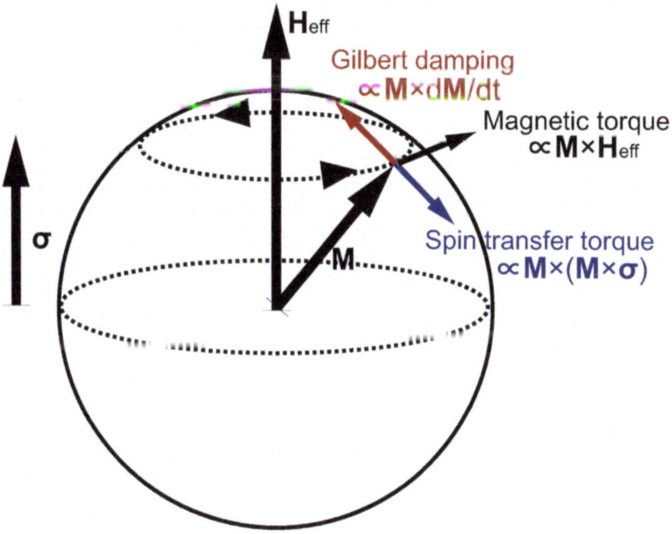

Fig. 1.3 A schematic illustration of the magnetic torques on the magnetization **M** under the effective magnetic field $\mathbf{H}_{\rm eff}$. The magnetic torque by $\mathbf{H}_{\rm eff}$ (black arrow) induces the magnetization precession around $\mathbf{H}_{\rm eff}$. The damping torque (red arrow) acts to align **M** in the $\mathbf{H}_{\rm eff}$ direction. In the presence of the interaction between **M** and a spin current with the spin angular momentum $\boldsymbol{\sigma}$ ($-\boldsymbol{\sigma}$), the spin transfer torque acts on **M** and competes (cooperate) with the damping torque

that acts on **M** in $\mathbf{H}_{\rm eff}$, which induces the magnetization precession around $\mathbf{H}_{\rm eff}$ (black arrow in Fig. 1.3). The second term is the magnetic damping torque, leading to the dissipation of the magnetization precession to align **M** in the $\mathbf{H}_{\rm eff}$ direction (red arrow in Fig. 1.3).

1.3.4 Spin Transfer Torque

Spin transfer torque is a magnetic torque exerted on a spin system via the spin transfer between conduction electrons and spin systems [16, 17]. Here, we consider the interaction between conduction electrons, whose spin is polarized in the $\boldsymbol{\sigma}$ direction, and localized electrons responsible for magnetization **M**. Via the exchange coupling, the conduction electrons transfer their spin angular momentum to the localized electrons, and the spin transfer torque $\boldsymbol{\tau}_{\rm STT}$ is exerted on **M** as

$$\boldsymbol{\tau}_{\rm STT} \propto \mathbf{M} \times (\mathbf{M} \times \hat{\boldsymbol{\sigma}}), \tag{1.9}$$

where $\hat{\boldsymbol{\sigma}}$ is the unit vector along the $\boldsymbol{\sigma}$ direction. A typical system where the spin transfer torque plays an important role is a bilayer structure of a paramagnetic metal and a ferromagnet (Fig. 1.4). By using the spin Hall effect in the paramagnetic metal, a spin current $j_{\rm s}$ can be injected into the ferromagnet, and the injected spin current

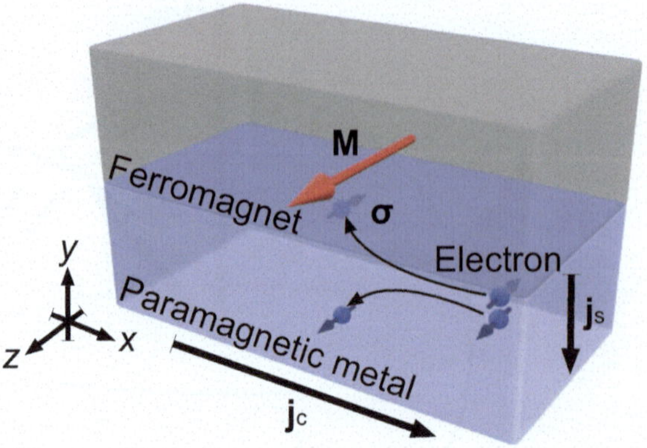

Fig. 1.4 A schematic illustration of the bilayer system of a ferromagnet and a paramagnetic metal. An electric current $\mathbf{j}_c$ is converted into a spin current $\mathbf{j}_s$ via the spin Hall effect in the paramagnetic metal, and the spin current is injected into the ferromagnet. The injected spin current exerts the spin transfer torque on the magnetization **M** in the ferromagnet

exerts the spin transfer torque $\boldsymbol{\tau}_{\mathrm{STT}}$ on the magnetization **M** in the ferromagnet. By taking $\boldsymbol{\tau}_{\mathrm{STT}}$ into account in the LLG equation, the magnetization dynamics in the ferromagnet is given by

$$\frac{d\mathbf{M}}{dt} = -\gamma \mathbf{M} \times \mathbf{H}_{\mathrm{ext}} + \frac{\alpha}{M_{\mathrm{s}}} \mathbf{M} \times \frac{d\mathbf{M}}{dt} + \frac{\gamma j_{\mathrm{s}}}{M_{\mathrm{s}}^2 d} \mathbf{M} \times (\mathbf{M} \times \hat{\boldsymbol{\sigma}}) \tag{1.10}$$

Here d is the thickness of the ferromagnet. As shown by the blue arrow in Fig. 1.3, the spin transfer torque competes with the damping torque and modulates the magnetic damping [18].

1.4 Lattice Dynamics

In this section, we explain the fundamental physics in elastic mechanics. We also introduce the mechanical dynamics of a mechanical oscillator.

1.4.1 Mechanics of Elasticity

Strain and stress tensor

As a starting point, we define a displacement tensor U as

$$U = \begin{pmatrix} \frac{\partial u_x}{\partial x} & \frac{\partial u_x}{\partial y} & \frac{\partial u_x}{\partial z} \\ \frac{\partial u_y}{\partial x} & \frac{\partial u_y}{\partial y} & \frac{\partial u_y}{\partial z} \\ \frac{\partial u_z}{\partial x} & \frac{\partial u_z}{\partial y} & \frac{\partial u_z}{\partial z} \end{pmatrix},$$

where (u_x, u_y, u_z) is the displacement vector at the position (x, y, z) in an elastic medium. We formulate the strain tensor in the elastic medium based on U. Fig. 1.5 shows a schematic illustration of the displacement (u_x, u_y) in the x-y plane, where

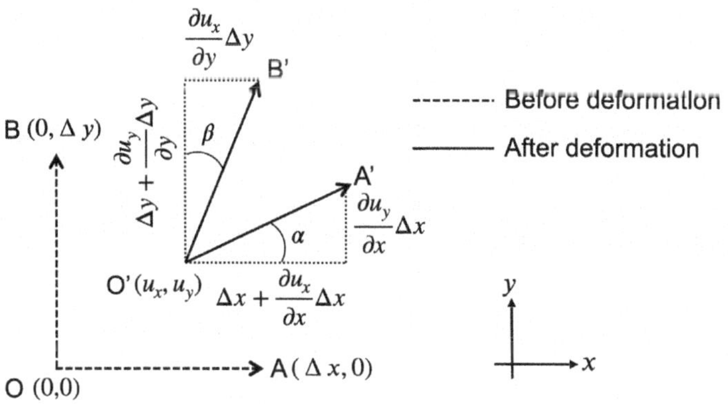

Fig. 1.5 A schematic illustration of the elastic deformation. The points A and B [position vectors $(\Delta x, 0)$ and $(0, \Delta y)$, respectively] change to the points A' and B' due to the displacement (u_x, u_y) in the x-y plane

the points A and B (position vectors $\overrightarrow{\mathrm{OA}}$ and $\overrightarrow{\mathrm{OB}}$, respectively) change to the points A' and B' (position vectors $\overrightarrow{\mathrm{O'A'}}$ and $\overrightarrow{\mathrm{O'B'}}$, respectively). In the presence of (u_x, u_y), $\overrightarrow{\mathrm{OA}}$ ($\overrightarrow{\mathrm{OB}}$) distorts along its length by $\partial u_x/\partial x \times \Delta x$ ($\partial u_y/\partial y \times \Delta y$), where Δx (Δy) is the original length of $\overrightarrow{\mathrm{OA}}$ ($\overrightarrow{\mathrm{OB}}$). As shown in Fig. 1.5, $\overrightarrow{\mathrm{OA}}$ ($\overrightarrow{\mathrm{OB}}$) also deforms in the transverse direction by $\partial u_y/\partial x \times \Delta x$ ($\partial u_x/\partial y \times \Delta y$). Note that these off-diagonal terms $\partial u_y/\partial x$ and $\partial u_x/\partial y$ in U consist of two different contributions of an elastic deformation [Fig. 1.6a] and a rigid body rotation [Fig. 1.6b]. To distinguish these components, we split U into the symmetric and asymmetric components as

$$U = \begin{pmatrix} \frac{\partial u_x}{\partial x} & \frac{1}{2}\left(\frac{\partial u_x}{\partial y} + \frac{\partial u_y}{\partial x}\right) & \frac{1}{2}\left(\frac{\partial u_x}{\partial z} + \frac{\partial u_z}{\partial x}\right) \\ \frac{1}{2}\left(\frac{\partial u_y}{\partial x} + \frac{\partial u_x}{\partial y}\right) & \frac{\partial u_y}{\partial y} & \frac{1}{2}\left(\frac{\partial u_y}{\partial z} + \frac{\partial u_z}{\partial y}\right) \\ \frac{1}{2}\left(\frac{\partial u_z}{\partial x} + \frac{\partial u_x}{\partial z}\right) & \frac{1}{2}\left(\frac{\partial u_z}{\partial y} + \frac{\partial u_y}{\partial z}\right) & \frac{\partial u_z}{\partial z} \end{pmatrix} + \begin{pmatrix} 0 & \frac{1}{2}\left(\frac{\partial u_x}{\partial y} - \frac{\partial u_y}{\partial x}\right) & \frac{1}{2}\left(\frac{\partial u_x}{\partial z} - \frac{\partial u_z}{\partial x}\right) \\ \frac{1}{2}\left(\frac{\partial u_y}{\partial x} - \frac{\partial u_x}{\partial y}\right) & 0 & \frac{1}{2}\left(\frac{\partial u_y}{\partial z} - \frac{\partial u_z}{\partial y}\right) \\ \frac{1}{2}\left(\frac{\partial u_z}{\partial x} - \frac{\partial u_x}{\partial z}\right) & \frac{1}{2}\left(\frac{\partial u_z}{\partial y} - \frac{\partial u_y}{\partial z}\right) & 0 \end{pmatrix} \quad (1.11)$$

The strain tensor e_{ij} is defined by the symmetric part (first term) of U in Eq. (1.11):

$$e_{ij} = \frac{1}{2}\left(\frac{\partial u_i}{\partial x_j} + \frac{\partial u_j}{\partial x_i}\right) \quad (1.12)$$

The antisymmetric part (second term) of U is called the rotation tensor

$$\omega_{ij} = \frac{1}{2}\left(\frac{\partial u_i}{\partial x_j} - \frac{\partial u_j}{\partial x_i}\right). \quad (1.13)$$

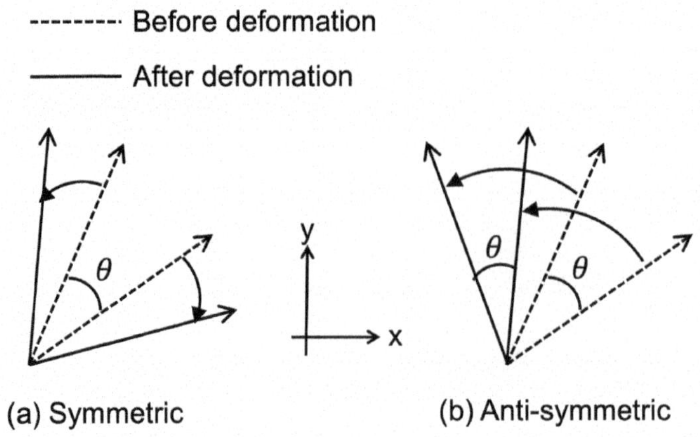

Fig. 1.6 A schematic illustration of the change in the vectors pointing at two positions before/after elastic distortion. **a** The elastic distortion that changes the inner product of the two vectors. **b** The elastic distortion without the change in the inner product of the two vectors

We consider the physical meaning of the strain tensor e_{ij} and the rotation tensor ω_{ij}. The diagonal component e_{ii} of the strain tensor corresponds to the distortion ratio in the i direction. Under the assumption $\partial u_i/\partial x_j \ll 1$, $\partial u_x/\partial y = \alpha$ and $\partial u_y/\partial x = \beta$, and the off-diagonal components is described as

$$e_{xy} = \frac{1}{2}\left(\frac{\partial u_x}{\partial y} + \frac{\partial u_y}{\partial x}\right) = \frac{1}{2}(\alpha + \beta), \tag{1.14}$$

where α and β are the distortion angle of the elastic body (Fig. 1.5). The result means that the off-diagonal component of the strain tensor corresponds to the angle change by the distortion in an elastic body, which is called a shear strain. By similar calculation the rotation tensor w_{ij} is given by

$$\omega_{xy} = \frac{1}{2}\left(\frac{\partial u_x}{\partial y} - \frac{\partial u_y}{\partial x}\right) = \frac{1}{2}(\beta - \alpha) \tag{1.15}$$

In the case of $\beta = -\alpha$ (Fig. 1.7), we can derive

$$e_{xy} = 0, \quad \omega_{xy} = -\alpha, \tag{1.16}$$

which means that w_{ij} is the angle of the rigid-body rotation of an elastic body.

We define the stress tensor and explain its properties. When an external force is applied to an elastic body, a strain is generated inside the elastic body, which induces an internal force. Stress is defined as the internal force per unit area acting on the surface of the elastic body. As shown in Fig. 1.8a, the stress acting on the surface can

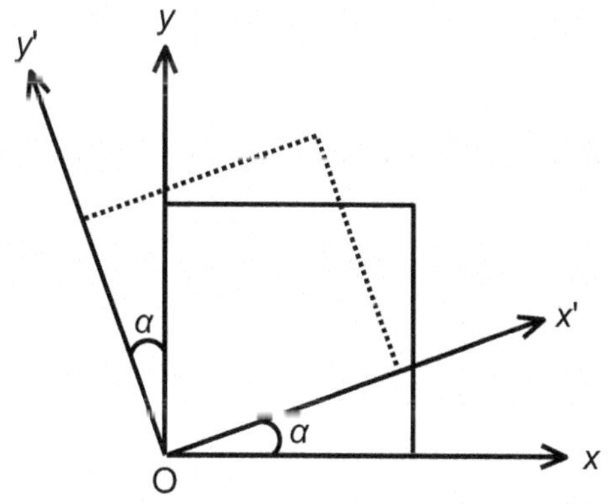

Fig. 1.7 A schematic illustration of the rigid-body rotation. α denotes the rotation angle

be described by two orthogonal vectors within the surface and one vector normal to the surface, which are called shear stress and normal stress, respectively. The nine stress components in Fig. 1.8a are described by the stress tensor

$$\sigma = \begin{pmatrix} \sigma_{xx} & \sigma_{xy} & \sigma_{xz} \\ \sigma_{yx} & \sigma_{yy} & \sigma_{yz} \\ \sigma_{zx} & \sigma_{zy} & \sigma_{zz} \end{pmatrix}. \tag{1.17}$$

The stress tensor becomes a symmetric tensor ($\sigma_{ij} = \sigma_{ji}$) from the condition that all torques are balanced in an elastic body. As shown in Fig. 1.8b, we consider a minute piece of the elastic body with the size $dx \times dy \times dz$ and calculate the torques around

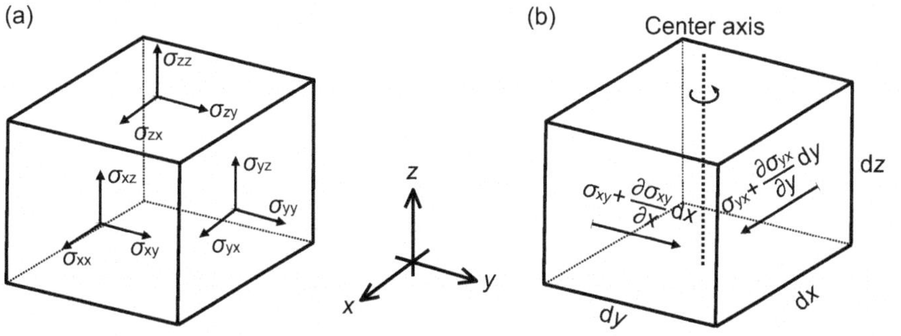

Fig. 1.8 **a** A schematic illustration of each component of the stress tensor σ_{ij} **b** A schematic illustration of the torques around the center of an elastic body in the z direction

its center axis. By imposing the condition that the torque along the z-axis is zero, we can obtain

$$\sigma_{xy} dydz \times \frac{dx}{2} + \left(\sigma_{xy} + \frac{\partial \sigma_{xy}}{\partial x} dx\right) dydz \times \frac{dx}{2}$$
$$-\sigma_{yx} dzdx \times \frac{dy}{2} + \left(\sigma_{yx} + \frac{\partial \sigma_{yx}}{\partial y} dy\right) dzdx \times \frac{dy}{2} = 0. \tag{1.18}$$

When we assume that the higher order terms of dx_i are negligibly small, we can derive

$$\sigma_{xy} = \sigma_{yx} \tag{1.19}$$

By calculating the torques along the x and y axes, we can obtain

$$\sigma_{yz} = \sigma_{zy}, \quad \sigma_{zx} = \sigma_{xz}, \tag{1.20}$$

which shows that the stress tensor satisfies

$$\sigma_{ij} = \sigma_{ji}. \tag{1.21}$$

In the linear response regime, the relation between σ_{ij} and e_{ij} is given by Hooke's law:

$$\sigma_{ij} = C_{ijkl} e_{kl}. \tag{1.22}$$

Here, the Einstein summation convention is used. In general, the constant C_{ijkl} has 81 components since it is the fourth order tensor, but the number of the components decreases due to the symmetry of the strain and stress tensors as

$$\sigma_{ij} = \sigma_{ji} \quad \Rightarrow \quad C_{ijkl} = C_{jikl} \tag{1.23}$$

$$e_{kl} = e_{lk} \quad \Rightarrow \quad C_{ijkl} = C_{ijlk} \tag{1.24}$$

In the following, we use the Voigt notation

$$\sigma_{xx} = \sigma_1, \quad \sigma_{yy} = \sigma_2, \quad \sigma_{zz} = \sigma_3, \quad \sigma_{yz} = \sigma_4, \quad \sigma_{zx} = \sigma_5, \quad \sigma_{xy} = \sigma_6 \tag{1.25}$$

$$e_{xx} = e_1, \quad e_{yy} = e_2, \quad e_{zz} = e_3, \quad 2e_{yz} = e_4, \quad 2e_{zx} = e_5, \quad 2e_{xy} = e_6, \tag{1.26}$$

and describe Hooke's law as

$$\sigma_i = c_{ij} e_j \quad (i, j = 1 \sim 6) \tag{1.27}$$

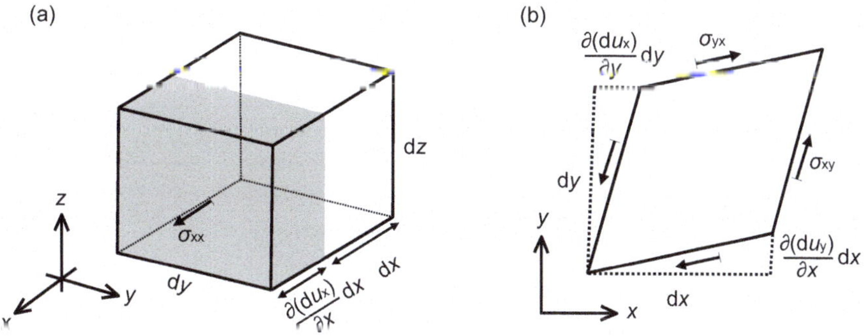

Fig. 1.9 **a** A schematic illustration of the normal stress σ_{xx} applied to an elastic body and the resultant strain (gray shaded area). **b** A schematic illustration of the shear stresses σ_{xy} and σ_{yx} applied to the elastic body and the resultant strain. The size of the elastic body is $dx \times dy \times dz$. u_x and u_y denote the displacement in the x and y directions, respectively

The coefficient c_{ij} is called the elastic constant, which has 36 components. If the crystal structure of materials is invariant with respect to a coordinate transformation, the number of independent components of the elastic constant is further reduced. For example, in the case of a cubic lattice, there are only three independent components of the elastic constant due to the 4-fold rotational symmetry along the x, y, and z axes as

$$\begin{pmatrix} \sigma_1 \\ \sigma_2 \\ \sigma_3 \\ \sigma_4 \\ \sigma_5 \\ \sigma_6 \end{pmatrix} = \begin{pmatrix} c_{11} & c_{12} & c_{12} & 0 & 0 & 0 \\ c_{12} & c_{11} & c_{12} & 0 & 0 & 0 \\ c_{12} & c_{12} & c_{11} & 0 & 0 & 0 \\ 0 & 0 & 0 & c_{44} & 0 & 0 \\ 0 & 0 & 0 & 0 & c_{44} & 0 \\ 0 & 0 & 0 & 0 & 0 & c_{44} \end{pmatrix} \begin{pmatrix} e_1 \\ e_2 \\ e_3 \\ e_4 \\ e_5 \\ e_6 \end{pmatrix}. \tag{1.28}$$

Elastic energy

The work acted on an elastic body by an external force is stored as the elastic energy via the deformation of the elastic body. First, we calculate the work dW'_{xx} that the normal stress σ_{xx} performs to the elastic body [Fig. 1.9a]. When we suppose that the elastic body distorts due to σ_{xx} by $de_{xx} \times dx = \partial(du_x)/\partial x \times dx$ in the x direction, where du_x is the displacement along the x-axis [grey shaded area in Fig. 1.9a], we can calculate dW'_{xx} as

$$dW'_{xx} = \sigma_{xx} \times dydz \times de_{xx}dx = \sigma_{xx}de_{xx}\Delta V, \tag{1.29}$$

where $\Delta V = dxdydz$. Then we can obtain the energy dW_{xx} per unit volume stored inside the elastic body as

$$dW_{xx} = \sigma_{xx}de_{xx} \tag{1.30}$$

Next, we calculate the work dW'_{xy} that the shear stresses σ_{xy} and σ_{yx} perform to the elastic body [Fig. 1.9b]. We define the strains due to σ_{xy} and σ_{yx} as de_{xy} and de_{yx}, respectively. As shown in Fig. 1.9b, we can obtain

$$dW'_{xy} = \sigma_{xy} \times dydz \times \frac{\partial(du_y)}{\partial x}dx + \sigma_{xy} \times dxdz \times \frac{\partial(du_x)}{\partial y}dy \quad (1.31)$$

$$= \sigma_{xy} \times \left(\frac{\partial(du_y)}{\partial x} + \frac{\partial(du_x)}{\partial y}\right) dxdydz, \quad (1.32)$$

$$= 2\sigma_{xy} de_{xy} \Delta V \quad (1.33)$$

where du_x (du_y) is the displacement along the x-axis (y-axis). We used $de_{xy} = de_{yx} = 1/2\left(\partial(du_y)/\partial x + \partial(du_x)/\partial y\right)$. The energy dW_{xy} per unit volume is given by

$$dW_{xy} = 2\sigma_{xy} de_{xy}. \quad (1.34)$$

We can calculate the other components in a similar way, and the total energy dW per unit volume is given by

$$\begin{aligned} dW &= \sigma_{xx} de_{xx} + \sigma_{yy} de_{yy} + \sigma_{zz} de_{zz} + 2\sigma_{xy} de_{xy} + 2\sigma_{yz} de_{yz} + 2\sigma_{zx} de_{zx} \\ &= \sigma_i de_i \\ &= c_{ij} e_j de_i. \end{aligned} \quad (1.35)$$

Note that we used the Voigt notation [Eqs. (1.25) and (1.26)]. In a system where the elastic constant is symmetric $c_{ij} = c_{ji}$, we can obtain the elastic energy W_{el} per unit volume by integrating the above equation:

$$W_{\text{el}} = \frac{1}{2} c_{ij} e_i e_j. \quad (1.36)$$

1.4.2 Mechanical Motion of Mechanical Oscillator

A micro mechanical oscillator such as a cantilever exhibits mechanical vibration with elastic deformation[19]. Its mechanical motion is a typical example of lattice dynamics based on elastic mechanics. In this section, starting from the elastic Hamiltonian of a mechanical oscillator, we derive the mechanical resonance mode of the mechanical vibration and explain its basic physics.

Equation of motion

We define the Hamiltonian of a mechanical oscillator as follows:

$$\mathcal{H} = \int_0^l \left(\frac{\Pi^2}{2\rho S} + \frac{EI_y}{2} \left(\frac{\partial^2 \delta}{\partial x^2} \right)^2 \right) dx \quad (1.37)$$

Here ρ, S, E, I_y, and l are the mass density, the cross-section area, the Young's modulus, the moment of inertia, and the length of the oscillator, respectively. $(\delta(x,t), \Pi(x,t))$ is the displacement and the momentum of the oscillator. The first (second) term corresponds to the kinetic (elastic) energy due to the mechanical vibration. For the second term, we assume that the length of the resonator does not change and consider only the geometrical deformation. From the Hamilton's canonical equations

$$\frac{d\delta(x)}{dt} = \frac{\partial \mathcal{H}}{\partial \Pi}, \quad \frac{d\Pi(x)}{dt} = -\frac{\partial \mathcal{H}}{\partial \delta}, \tag{1.38}$$

we can obtain the equation of motion on $\delta(x,t)$ as

$$\frac{Ed^2}{12}\frac{\partial^4 \delta}{\partial x^4} + \rho \frac{\partial^2 \delta}{\partial t^2} = 0 \tag{1.39}$$

Here we used $S = wd$, $I_y = d^3 w/12$ by assuming the rectangular shape of the mechanical oscillator. w and d are the width and the thickness of the oscillator, respectively.

Resonant mechanical vibration mode

We derive the resonance mode of the mechanical vibration from the equation of motion. By assuming that the solution is variable-separable with respect to position and time, $\delta(x,t) = z(x)e^{-i\omega t}$, we can obtain the differential equation for $z(x)$ as

$$\frac{d^4 z(x)}{dx^4} = k^4 z(x), \tag{1.40}$$

where k satisfies

$$\omega = k^2 d\sqrt{\frac{E}{12\rho}}. \tag{1.41}$$

By solving Eq. (1.40), we can describe $z(x)$ with four arbitrary constants as

$$z(x) = A\sin(kx) + B\cos(kx) + C\sinh(kx) + D\cosh(kx). \tag{1.42}$$

Equation (1.42) is a general solution of a resonant mechanical vibration of a mechanical oscillator. As can be seen from the k dependence of the solution, the constant k corresponds to the wave number of the resonant mechanical vibration, meaning that Eq. (1.41) can be interpret as the dispersion relation of the vibration. By using the above solution, we calculate the resonant mechanical vibration of a bridge structure shown in Fig. 1.10a. Here, we impose the following four boundary conditions for the bridge structure:

$$z(0) = 0, \quad z'(0) = 0, \quad z(l) = 0, \quad z'(l) = 0, \tag{1.43}$$

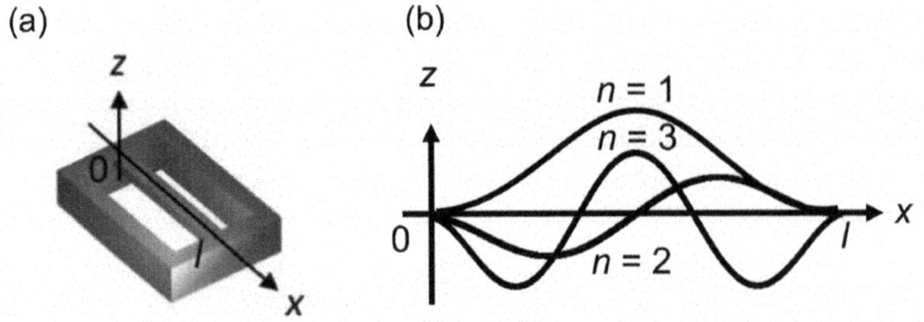

Fig. 1.10 **a** A schematic illustration of a micro mechanical oscillator. l denotes the length of the oscillator. **b** The spatial profile of the nth resonant mechanical vibration of the oscillator

from which we can determine the condition that the wavenumber k should satisfy as

$$\cos(k_n l)\cosh(k_n l) = 1 \tag{1.44}$$

$$k_n = \frac{a_n}{l}, \quad a_n = 4.730,\ 7.853\ ,, \tag{1.45}$$

As shown in Fig. 1.10b, the discrete value of k_n corresponds to the wave number of the nth resonant mechanical vibration. By substituting Eq. (1.45) to Eq. (1.41), we can obtain the nth mechanical resonance frequency f_n as

$$f_n = \frac{\omega_n}{2\pi} = \frac{k_n^2 d}{2\pi}\sqrt{\frac{E}{12\rho}} = \frac{2\pi d}{\lambda_n^2}\sqrt{\frac{E}{12\rho}}. \tag{1.46}$$

Here, we defined $\lambda_n = 2\pi/k_n$ that corresponds to the wavelength of the resonant vibration.

1.5 Spin-Lattice Coupling

In this section, we derive the magnetoelastic energy based on the spin-lattice coupling and formulate the magnetostriction.

1.5.1 Magnetoelastic Energy

We consider the spin pair model [Fig. 1.11a] in which the spins are located on the crystal lattice. As a starting point, we define the energy w in the model with respect to the lattice distance r and the relative angle ϕ between the spins and the crystal lattice axis as

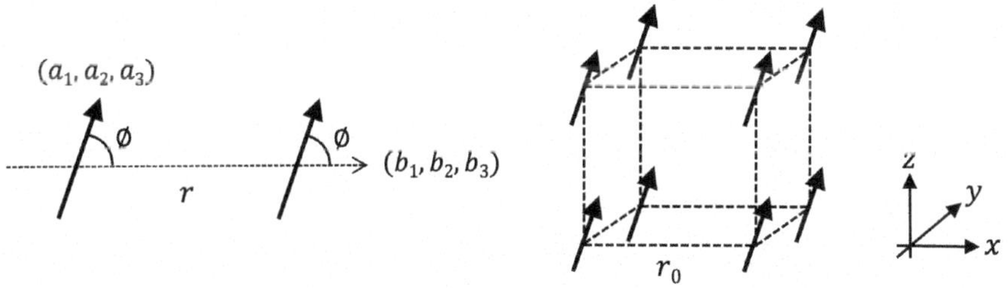

(a) A spin pair on lattices (b) Spins in a cubic crystal

Fig. 1.11 **a** A schematic illustration of a spin pair on a crystal axis. (a_1, a_2, a_3) and (b_1, b_2, b_3) are direction cosines of the spins and the crystal axis direction, respectively. The angle of the spins and the lattice distance are defined as ϕ and r, respectively. **b** A schematic illustration of spin pairs on a cubic lattice with the lattice distance r_0

$$w(r, \cos\phi) = l(r)\left(\cos^2\phi - \frac{1}{3}\right), \tag{1.47}$$

where the coefficient $l(r)$ depends on r. This equation shows that the spin orientation ϕ and the lattice distance r are correlated. By using the direction cosine of the spin (a_1, a_2, a_3) and the direction cosine connecting the spins (b_1, b_2, b_3), as shown in Fig. 1.11a, we can describe the energy $w(r, \cos\phi)$ of the spin pair as

$$w(r, \cos\phi) = l(r)\left((a_1 b_1 + a_2 b_2 + a_3 b_3)^2 - \frac{1}{3}\right). \tag{1.48}$$

In the following, for simplicity, we consider a cubic lattice [Fig. 1.11b], where r_0 is the distance between spins. For the spin pair along the x direction, we can obtain its energy by substituting $r = r_0$ and $(b_1, b_2, b_3) = (1, 0, 0)$ to Eq. (1.48) as

$$w(r_0, \cos\phi_0) = l(r_0)\left(a_1^2 - \frac{1}{3}\right). \tag{1.49}$$

We then calculate the change in $w(r_0, \cos\phi)$ when the crystal is distorted under the strain tensor e_{ij}. In the presence of the strain, the vector between the spins along the x direction $r_0(1, 0, 0)$ changes to $\approx r_0(1 + e_{xx}, e_{yx}, e_{zx})$ under the assumption that there is no rigid body rotation (Fig. 1.12). From the result, we can calculate the lattice distance in the x direction as

$$r_0\sqrt{(1 + e_{xx})^2 + e_{zx}^2 + e_{yx}^2} \approx r_0(1 + e_{xx}) \tag{1.50}$$

Here, we assumed that the second order or higher terms of the strain tensor are negligibly small. The leading term of the direction cosine is given by

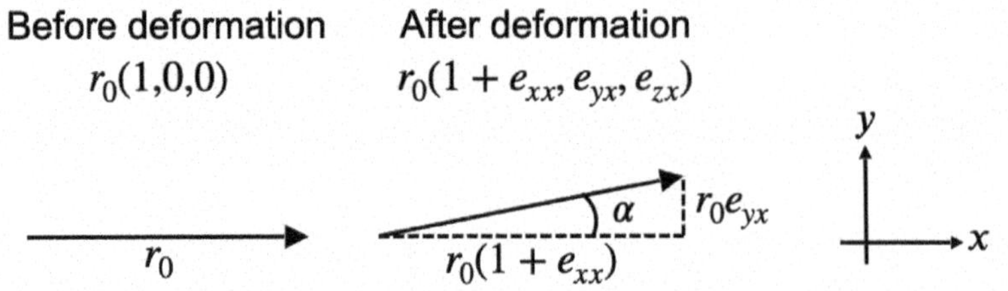

Fig. 1.12 A schematic illustration of the vector with the length r_0 before/after deformation. The strain e_{ij} changes r_0 and the angle α

$$(b_1, b_2, b_3) \approx \left(1,\ e_{yx},\ e_{zx}\right) \tag{1.51}$$

By substituting Eqs. (1.50) and (1.51) to Eq. (1.48) we can derive

$$w(r_0(1 + e_{xx}), \cos\phi) = l(r_0(1 + e_{xx}))\left((a_1 + a_2 e_{yx} + a_3 e_{zx})^2 - \frac{1}{3}\right) \tag{1.52}$$

$$= \left(l(r_0) + \left(\frac{\partial l}{\partial r}\right)_{r_0} r_0 e_{xx} + \mathrm{O}(e^2)\right)$$
$$\times \left((a_1 + a_2 e_{yx} + a_3 e_{zx})^2 - \frac{1}{3}\right) \tag{1.53}$$

$$= w(r_0, \cos\phi_0) + \left(\frac{\partial l}{\partial r}\right)_{r_0} r_0 \left(a_1^2 - \frac{1}{3}\right) e_{xx}$$
$$+ 2l(r_0)a_1 a_2 e_{xy} + 2l(r_0)a_3 a_1 e_{zx} + \mathrm{O}(e^2) \tag{1.54}$$
$$= w(r_0, \cos\phi) + \Delta w_x, \tag{1.55}$$

where

$$\Delta w_x = \left(\frac{\partial l}{\partial r}\right)_{r_0} r_0 \left(a_1^2 - \frac{1}{3}\right) e_{xx} + 2l(r_0)a_1 a_2 e_{xy} + 2l(r_0)a_3 a_1 e_{zx} \tag{1.56}$$

We can calculate the energy of the spin pair along the y and z direction in a similar way and obtain

$$\Delta w_y = \left(\frac{\partial l}{\partial r}\right)_{r_0} r_0 \left(a_2^2 - \frac{1}{3}\right) e_{yy} + 2l(r_0)a_2 a_3 e_{yz} + 2l(r_0)a_1 a_2 e_{xy} \tag{1.57}$$

$$\Delta w_z = \left(\frac{\partial l}{\partial r}\right)_{r_0} r_0 \left(a_3^2 - \frac{1}{3}\right) e_{zz} + 2l(r_0)a_3 a_1 e_{zx} + 2l(r_0)a_2 a_3 e_{yz} \tag{1.58}$$

By summing up the terms for all spin pairs per unit volume, we can calculate the magnetoelastic energy W_{me} as

$$W_{\mathrm{me}} = B_1\left(\left(a_1^2 - \frac{1}{3}\right)e_{xx} + \left(a_2^2 - \frac{1}{3}\right)e_{yy} + \left(a_3^2 - \frac{1}{3}\right)e_{zz}\right) + B_2(a_1a_2e_{xy} + a_2a_3e_{yz} + a_3a_1e_{zx}) \quad (1.59)$$

Here, N is the number of spin pairs along each crystal axis per unit volume, $B_1 = Nr_0\left(\frac{\partial l}{\partial r}\right)_{r_0}$, and $B_2 = 4Nl(r_0)$.

1.5.2 Magnetostriction

We calculate the magnetostriction e_{ij} by minimizing the sum of the magnetoelastic energy and the elastic energy $W = W_{\mathrm{me}} + W_{\mathrm{el}}$:

$$\frac{\partial W}{\partial e_{ij}} = 0 \quad (1.60)$$

Here the explicit form of the elastic energy W_{el} is given by [Eqs. (1.28) and (1.36)]

$$\begin{aligned} W_{\mathrm{el}} &= \frac{1}{2}c_{11}(e_1^2 + e_2^2 + e_3^2) + \frac{1}{2}c_{44}(e_4^2 + e_5^2 + e_6^2) \\ &+ c_{12}(e_1e_2 + e_2e_3 + e_3e_1) \quad (1.61) \\ &= \frac{1}{2}c_{11}(e_{xx}^2 + e_{xx}^2 + e_{xx}^2) + 2c_{44}(e_{xy}^2 + e_{yz}^2 + e_{zx}^2) \\ &+ c_{12}(e_{xx}e_{yy} + e_{yy}e_{zz} + e_{zz}e_{xx}) \quad (1.62) \end{aligned}$$

By solving the above local minimum condition we can obtain

$$e_{xx} = \frac{B_1}{c_{12} - c_{11}}\left(a_1^2 - \frac{1}{3}\right),\ e_{yy} = \frac{B_1}{c_{12} - c_{11}}\left(a_2^2 - \frac{1}{3}\right),\ e_{zz} = \frac{B_1}{c_{12} - c_{11}}\left(a_3^2 - \frac{1}{3}\right), \quad (1.63)$$

$$e_{xy} = -\frac{B_2}{4c_{44}}a_1a_2,\ e_{yz} = -\frac{B_2}{4c_{44}}a_2a_3,\ e_{zx} = -\frac{B_2}{4c_{44}}a_3a_1. \quad (1.64)$$

The result means that an elastic body exhibits the strain that is related to the magnetization direction (a_1, a_2, a_3). For example, when the magnetization is along the x axis $(a_1, a_2, a_3) = (1, 0, 0)$, the strain is given by

$$e_{xx} = \frac{2}{3}\frac{B_1}{c_{12} - c_{11}},\ e_{yy} = -\frac{1}{3}\frac{B_1}{c_{12} - c_{11}},\ e_{zz} = -\frac{1}{3}\frac{B_1}{c_{12} - c_{11}},\ e_{xy} = e_{yz} = e_{zx} = 0, \quad (1.65)$$

where the expansion in the x direction and the contraction in the y and z directions occur. For the general case, we obtain the distortion λ_{ms} along the (d_1, d_2, d_3) direction in an elastic body magnetized in the (a_1, a_2, a_3) direction. To calculate λ_{ms}, we consider the change in a unit vector $\mathbf{n} = (d_1, d_2, d_3)$ in the presence of the strain. When the unit vector changes $\mathbf{n} \to \mathbf{n'}$ due to the strain, as shown in Fig. 1.13, we can derive λ_{ms} as

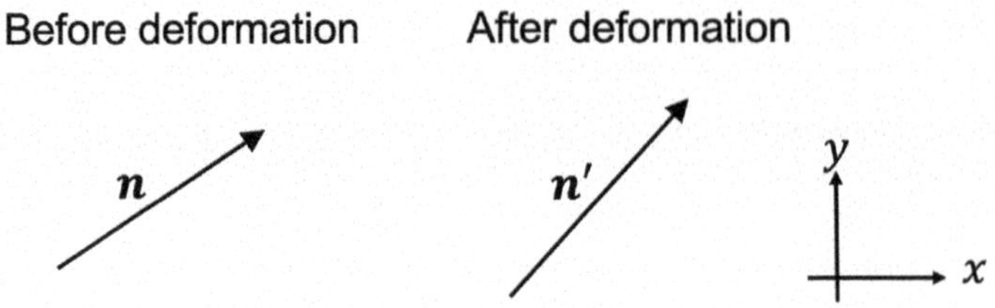

Fig. 1.13 A schematic illustration of the vectors **n** and **n'** before and after deformation, respectively

$$\lambda_{ms} = \frac{\mathbf{n} \cdot \mathbf{n'} - 1}{|\mathbf{n}|} \tag{1.66}$$

We can describe **n'** with respect to the strain tensor [Eqs. (1.63) and (1.64)] and $\mathbf{n} = (d_1, d_2, d_3)$ in a similar way as shown in Fig. 1.12 as

$$\lambda_{ms} = \frac{B_1}{c_{12} - c_{11}} \left(a_1^2 d_1^2 + a_2^2 d_2^2 + a_3^2 d_3^2 - \frac{1}{3} \right) - \frac{B_2}{2c_{44}} (a_1 a_2 d_1 d_2 + a_2 a_3 d_2 d_3 + a_3 a_1 d_3 d_1) \tag{1.67}$$

From the above equation, we calculate the magnetostriction λ_{100} (λ_{111}) in the $(d_1, d_2, d_3) = (1, 0, 0)$ direction [$(d_1, d_2, d_3) = (\frac{1}{\sqrt{3}}, \frac{1}{\sqrt{3}}, \frac{1}{\sqrt{3}})$ direction] when the magnetization is along the same (d_1, d_2, d_3) direction:

$$\lambda_{100} = \frac{2}{3} \frac{B_1}{c_{12} - c_{11}}. \tag{1.68}$$

$$\lambda_{111} = -\frac{1}{6} \frac{B_2}{c_{44}} \tag{1.69}$$

By using Eqs. (1.68) and (1.69), λ_{ms} is given by

$$\lambda_{ms} = \frac{3}{2} \lambda_{100} \left(a_1^2 d_1^2 + a_2^2 d_2^2 + a_3^2 d_3^2 - \frac{1}{3} \right) + 3\lambda_{111} (a_1 a_2 d_1 d_2 + a_2 a_3 d_2 d_3 + a_3 a_1 d_3 d_1) \tag{1.70}$$

In particular, in the isotropic case

$$\lambda_{100} = \lambda_{111} = \lambda, \tag{1.71}$$

the magnetostriction is simplified to

$$\lambda_{ms} = \frac{3}{2} \lambda \left((a_1 d_1 + a_2 d_2 + a_3 d_3)^2 - \frac{1}{3} \right) = \frac{3}{2} \lambda \left(\cos^2\theta - \frac{1}{3} \right). \tag{1.72}$$

Here θ is the relative angle between the magnetization direction (a_1, a_2, a_3) and the (d_1, d_2, d_3) direction.

References

1. Kato YK, Myers RC, Gossard AC, Awschalom DD (2004) Science 306:1910
2. Saitoh E, Ueda M, Miyajima H, Tatara G (2006) Appl Phys Lett 88:182509
3. Uchida K, Takahashi S, Harii K, Ieda J, Koshibae W, Ando K, Maekawa S, Saitoh E (2008) Nature 455:778
4. Flipse J, Dejene FK, Wagenaar D, Bauer GEW, Ben Youssef J, van Wees BJ (2014) Phys Rev Lett 113, 027601
5. Matsuo M, Saitoh E, Maekawa S (2017) J Phys Soc Jpn 86:011011
6. Takahashi R, Matsuo M, Ono M, Harii K, Chudo H, Okayasu S, Ieda J, Takahashi S, Maekawa S, Saitoh E (2016) Nat Phys 12:52
7. Wallis TM, Moreland J, Kabos P (2006) Appl Phys Lett 89:122502
8. Losby JE, Sani FF, Grandmont DT, Diao Z, Belov M, Burgess JAJ, Compton SR, Hiebert WK, Vick D, Mohammad K, Salimi E, Bridges GE, Thomson DJ, Freeman MR (2015) Science 350:798
9. Nguyen CTC, Howe RT (1999) IEEE J Solid State Circuits 34:440
10. Oikawa Y, Arisawa H, Daimon S, Saitoh E (2018) Appl Phys Lett 113:142407
11. Materials ferromagnetic. In: Wohlfarth EP. North-Holland Publishing Company (1980)
12. Arisawa H, Daimon S, Oikawa Y, Seo Y-J, Harii K, Oyanagi K, Saitoh E (2019) Appl Phys Lett 114:122402
13. Arisawa H, Shim H, Daimon S, Kikkawa T, Oikawa Y, Takahashi S, Ono T, Saitoh E (2022) Nat Commun 13:2440
14. Hoffmann A (2013) IEEE Trans Magn 49:5172
15. Spin waves. In: Stancil DD, Prabhakar A. Springer (2009)
16. Slonczewski JC (1996) J Magn Magn Mater 159:L1
17. Berger L (1996) Phys Rev B 54:9353
18. Ando K, Takahashi S, Harii K, Sasage K, Ieda J, Maekawa S, Saitoh E (2008) Phys Rev Lett 101:036601
19. Foundations of nanomechanics: from solid-state theory to device applications. In: Cleland AN. Springer, New York (2003)

Chapter 2
Experimental Methods

In the present study, we constructed a mechanical displacement measurement system using a laser Doppler interferometer aiming for highly sensitive detection of the magnetomechanical response. In this section, we explain the physical properties of the samples used in the experiment, the sample preparation method, and the measurement method.

2.1 Physical Properties of Samples

2.1.1 Ferrimagnetic Insulator $Y_3Fe_5O_{12}$

In the first study on the correlation between the magnetization and the mechanical vibration in a micro magnetic mechanical oscillator, we used a ferrimagnetic insulator yttrium iron garnet $Y_3Fe_5O_{12}$ (YIG). YIG exhibits negligibly small crystalline magnetic anisotropy, and its magnetization process is dominated by the shape magnetic anisotropy, which is determined by the sample shape. The characteristic enables us to control the magnetic anisotropy of the YIG mechanical oscillator by using microfabrication technologies. YIG is also considered to be one of the most suitable materials for studying spin dynamics [1, 2] because of its small magnetic damping. In other words, the YIG mechanical oscillators could be a good playground for the spin-lattice coupled dynamics including spin waves.

H. Arisawa, *A Study on Spin Mechanics Phenomena in Spin-Lattice Coupled Systems*, Springer Theses, https://doi.org/10.1007/978-981-97-8642-8_2

2.1.2 Ferromagnet $Tb_{0.3}Dy_{0.7}Fe_2$

In the second research, we tried to demonstrate a spin-current induced volume modulation effect in a giant magnetostrictive material, terbium dysprosium iron $Tb_{0.3}Dy_{0.7}Fe_2$. $Tb_{0.3}Dy_{0.7}Fe_2$ is an alloy system of Fe and rare earth elements Tb and Dy, which exhibits ferromagnetic order at room temperature. Its magnetic properties are summarized in Table 2.1. The ferromagnetic transition temperature, the saturation magnetic field, and the coercive field are about 670 K, 2000 Oe, and 60 Oe, respectively. The remarkable property is that the magnetostriction of $Tb_{0.3}Dy_{0.7}Fe_2$ is much greater than that of typical ferromagnets. The magnetostriction of $Tb_{0.3}Dy_{0.7}Fe_2$ is $\approx$ 1700 ppm [3], which is two orders larger than that of permalloy $\approx$ 10 ppm. For this reason, $Tb_{0.3}Dy_{0.7}Fe_2$ is widely used for a magnetic actuator.

The origin of the giant magnetoelastic coupling in $Tb_{0.3}Dy_{0.7}Fe_2$ is associated with the structural phase transition at the morphotropic phase boundary (MPB) [6]. MPB refers to a phase boundary where the crystal structure changes depending on temperature and chemical composition. Figure 2.1a shows the phase diagram of $Tb_{1-x}Dy_xFe_2$ with respect to temperature T and the composition ratio x. Recent studies have confirmed that the crystal structure changes from rhombohedral to tetragonal at around 200 K$< T <$300 K and $0.7 < x < 0.8$ by performing the X-ray crystal-structure analysis. Figure 2.1b shows a two-dimensional X-ray intensity map with respect to temperature and d-spacing. At $T \approx 240$ K, the X-ray intensity changes from a single peak to double peaks, which corresponds to the crystal structure transition from tetragonal to rhombohedral. Moreover, it is known that the magnetic anisotropy changes from the (001) direction to the (111) direction at this phase boundary concomitant to the crystal structure change. In the previous study, it is confirmed that the slope of the M-T curve changes at the phase boundary at $T \approx 240$ K, which reflects the magnetic anisotropic transition [white triangles in Fig. 2.1b]. As is in the same analogy with the MPB in ferroelectric materials [7], the presence of the MPB at around room temperature is thought to cause the large magnetoelastic response in $Tb_{0.3}Dy_{0.7}Fe_2$.

Table 2.1 Magnetic properties of a ferromagnet $Tb_{0.3}Dy_{0.7}Fe_2$

	Magnetic properties of $Tb_{0.3}Dy_{0.7}Fe_2$
Magnetic transition temperature	670 K [4]
Saturation field	2000 Oe [5]
Coercive field	60 Oe [3]
Magnetostriction	1700 ppm [3]

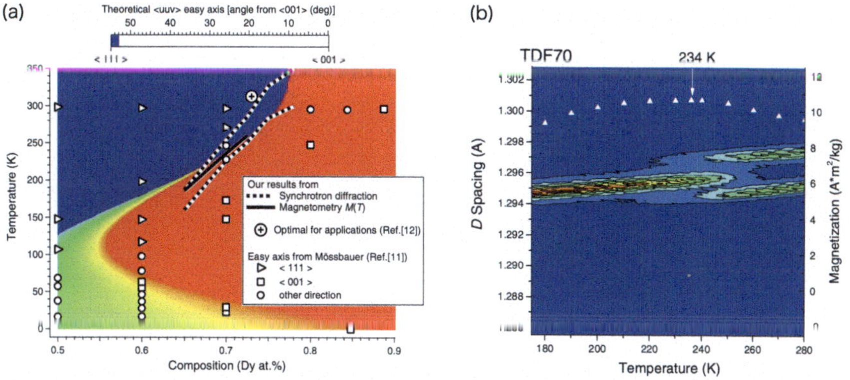

Fig. 2.1 **a** Phase diagram of $Tb_{1-x}Dy_xFe_2$ with respect to temperature and the composition ratio x. **b** The X-ray intensity color map with respect to temperature and d-spacing. The rhombohedral-tetragonal structural phase transition occurs at the temperature at which the X-ray intensity changes from a single peak to double peaks. The magnetization takes the local maximum at the structural phase transition temperature (white triangles). Adapted from Phys. Rev. Lett. **111**, 017203 (2013) [6]; copyright 2013 American Physical Society

2.2 Sample Fabrication

2.2.1 Focused Ion Bean Method

We used the focused ion beam (FIB) method to fabricate a YIG mechanical oscillator. In the FIB method, as shown in Fig. 2.2, a Ga^+ ion beam is focused onto the sample to knock out the sample atoms, enabling microfabrication into a desired structure. First, Ga^+, which is ionized in high vacuum by high voltage, is irradiated from an ion gun. Ga^+ is converged and accelerated into an ion beam by a magnetic lens, an aperture, and a condenser, and it collides on the sample surface. The irradiated Ga^+ ions give their kinetic energy to the sample surface, and when the kinetic energy exceeds the binding energy of the atoms, they are removed from the surface. Since the alignment

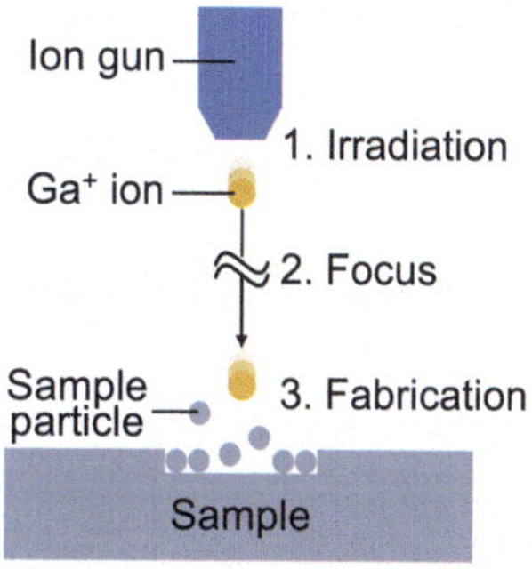

Fig. 2.2 A schematic illustration of microfabrication by the focused ion beam (FIB) method. Ga^+ irradiated from an ion gun is converged and accelerated into an ion beam and the Ga^+ ion beam is focused onto the sample surface to knock out the sample atoms

of the Ga^+ beam is highly controllable by modulating the accelerating voltage, precise microfabrication is possible for many kinds of materials including YIG [8]. In the experiment, we used the FIB device (Versa3D Dual Beam) manufactured by FEI Company.

2.2.2 *Electroplating Method*

In the second study, we performed electroplating to fabricate the $Tb_{0.3}Dy_{0.7}Fe_2$/paramagnetic-metal bilayer system [9, 10]. By using the electroplating method, it is possible to deposit a magnetostrictive $Tb_{0.3}Dy_{0.7}Fe_2$ film on a paramagnetic metal. The electroplating process was carried out with the cooperation of Professor Takahito Ono and a doctoral student Hang Shim in the Graduate School of Engineering, Tohoku University. First, a paramagnetic metal was deposited on a Si substrate by RF sputtering. Basic information on the deposited paramagnetic metals is shown in Table 2.2. Ar gas was introduced during the sputtering, and the pressure during the film deposition is about 10^{-1} Pa.

Next, as shown in Fig. 2.3, we prepared a solution containing Tb, Dy, and Fe ions, and placed the (Pt, W, or Cu)/Si substrate and a Pt wire in the solution as a working electrode and a counter electrode, respectively. To monitor the voltage between the two electrodes, we prepared a reference electrode, a AgCl wire, in a KCl solution, which is electrically connected to the solution for the working and counter electrodes with a KCl salt bridge. By adjusting the voltage and deposition time at the electrolyte temperature 40 °C, where the typical deposition rate is 100 nm/hour, we fabricated the $Tb_{0.3}Dy_{0.7}Fe_2$ film on the (Pt, W, or Cu)/Si substrate. The advantage of the above electroplating method is that we can obtain a $Tb_{0.3}Dy_{0.7}Fe_2$ film with the large magnetostriction compared to the sputtering method [11].

Table 2.2 Basic information on the deposited paramagnetic films

Sample	Thickness	Deposition method
Pt	140 nm	RF sputtering
W	100 nm	RF sputtering
Cu	100 nm	RF sputtering

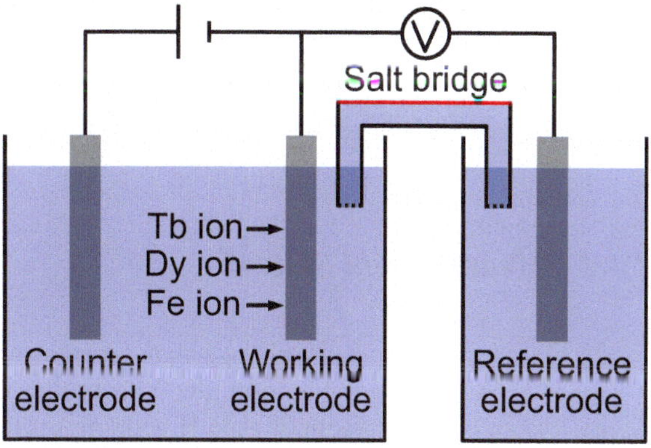

Fig. 2.3 A schematic illustration of the electroplating deposition of the $Tb_{0.3}Dy_{0.7}Fe_2$ film on the paramagnetic metal. Reproduced from Nat. Commun. **13**, 2440 (2022) [10]; copyright 2022 Author(s), licensed under a Creative Commons Attribution (CC BY) license

2.3 Mechanical Response Measurement with Laser Doppler Vibrometer

2.3.1 Laser Doppler Vibrometry

To sensitively detect the magnetomechanical response, we used a laser Doppler vibrometer. Figure 2.4 shows a schematic illustration of the operating principle. The basic components are a laser light source, a beam splitter, and a photodetector that detects light interference as a voltage. First, a laser (wavelength $\lambda = 532$ nm) emitted from the laser source is split into two laser beams, a reference beam and an incident beam, by a beam splitter. The incident laser beam is focused on the surface of the sample by an objective lens, and its reflected component joins the reference beam via the beam splitter. Here, the laser reflected from the sample surface undergoes frequency modulation $v(t)/\lambda$ reflecting the velocity $v(t)$ of the sample surface due to Doppler effect, which results in the interference between the reference beam and

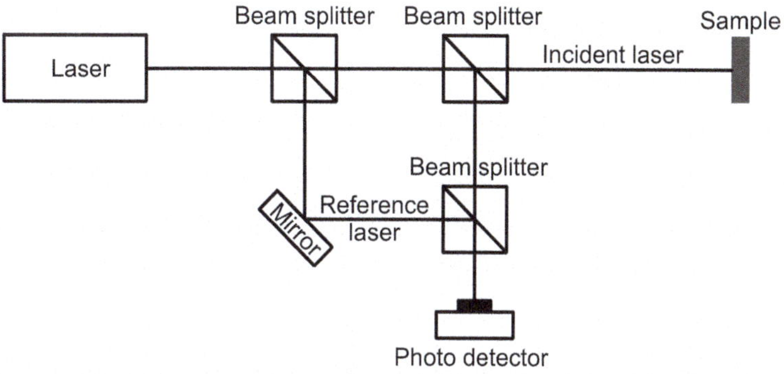

Fig. 2.4 A schematic illustration of the configuration of the laser Doppler vibrometer

the reflected beam. By detecting the frequency of the interference laser light with the photodetector, the sample surface velocity $v(t)$ is measured. The mechanical displacement $x(t)$ is obtained from the time integration of $v(t)$ and the Fourier spectrum $\tilde{x}(\omega)$ is obtained by the Fourier transform of $x(t)$.

2.3.2 Measurement System

In the first study, we measured the magnetomechanical response of a micro YIG bridge [12]. As shown in Fig. 2.5a, the YIG bridge was placed on a piezoelectric device and an in-plane magnetic field **H** was applied to the YIG bridge. As an input, we applied an a.c. voltage at the frequency f_P to a piezoelectric device and induced the longitudinal vibration of the piezoelectric device. The longitudinal vibration of the piezoelectric device excites the mechanical vibration of the YIG bridge. As an output, to investigate the correlation between the magnetization and the mechanical

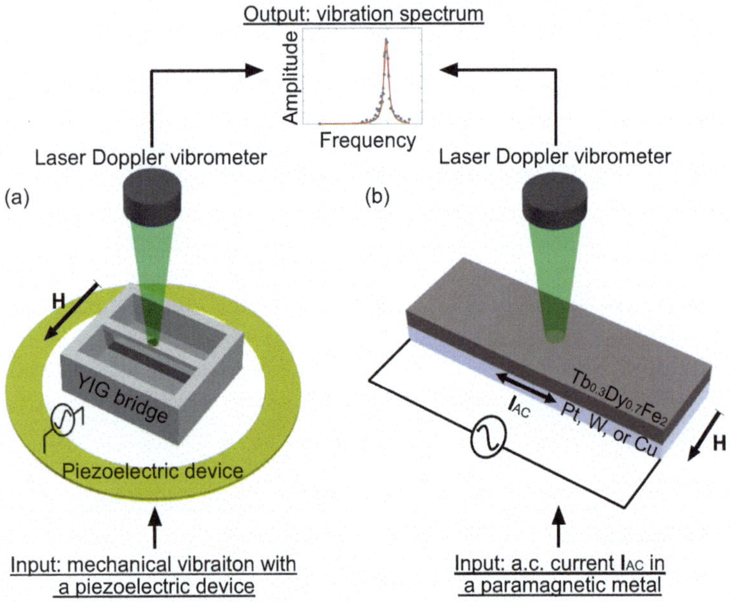

Fig. 2.5 Schematic illustrations of the measurement system. **a** Mechanical vibration measurement in a YIG bridge under the magnetic field **H**. By applying an a.c. voltage to a piezoelectric device, the longitudinal vibration of the piezoelectric device was induced. The mechanical vibration spectrum of the bridge was measured with a laser Doppler vibrometer. **b** Mechanical vibration measurement of a $Tb_{0.3}Dy_{0.7}Fe_2$ film in response to the spin current injection. An a.c. spin current was injected into the $Tb_{0.3}Dy_{0.7}Fe_2$ film by applying an a.c. electric current $\mathbf{I}_{AC}$ with the frequency f_{AC} in the paramagnetic metal layer attached to the $Tb_{0.3}Dy_{0.7}Fe_2$ film. The mechanical vibration spectrum of the $Tb_{0.3}Dy_{0.7}Fe_2$ film was measured with a laser Doppler vibrometer

vibration in the bridge, we measured the out-of-plane mechanical vibration spectrum of the YIG bridge under **H** with the laser Doppler vibrometer.

In the second study, we injected a spin current into a $Tb_{0.3}Dy_{0.7}Fe_2$ film and measured its mechanical response to the injected spin current [Fig. 2.5b] [10]. An a.c. current $\mathbf{I}_{AC}$ with the frequency f_{AC} was applied to the paramagnetic metal (Pt, W, or Cu) film that is attached to the $Tb_{0.3}Dy_{0.7}Fe_2$ film. Via the spin Hall effect, an a.c. spin current is injected into the $Tb_{0.3}Dy_{0.7}Fe_2$ film. We measured the out-of-plane mechanical vibration spectrum of the $Tb_{0.3}Dy_{0.7}Fe_2$ film in the magnetic field **H**. In the actual measurement, the $Tb_{0.3}Dy_{0.7}Fe_2$/(Pt, W, or Cu) sample was cut into a rectangular shape of $\approx 2\,\text{mm} \times 10\,\text{mm}$. All measurements were performed in a vacuum of about 10^{-4} Pa.

References

1. Stancil DD, Prabhakar A (ed) (2009) Spin waves
2. Arisawa H, Daimon S, Oikawa Y, Kikkawa T, Saitoh E (2023) Phys Rev B 107:134408
3. Sandlund L, Fahlander M, Cedell T, Clark AE, Restorff JB, Wun-Fogle M (1994) J Appl Phys 75:5656
4. Wang BL, Jin YM (2012) J Appl Phys 111:103908
5. Clark AE, Abbundi R, Gillmor WR (1978) IEEE Trans Magn MAG-14, 54
6. Bergstrom R Jr, Wuttig M, Cullen J, Zavalij P, Briber R, Dennis C, Garlea VO, Laver M (2013) Phys Rev Lett 111:017203
7. Yamashita Y (1994) Jpn J Appl Phys 33:5328
8. Oikawa Y, Arisawa H, Daimon S, Saitoh E (2018) Appl Phys Lett 113:142407
9. Shim H, Sakamoto K, Inomata N, Toda M, Toan NV, Ono T (2020) Micromachines 11:523
10. Arisawa H, Shim H, Daimon S, Kikkawa T, Oikawa Y, Takahashi S, Ono T, Saitoh E (2022) Nat Commun 13:2440
11. Panduranga MK, Lee T, Chavez A, Prikhodko SV, Carman GP (2018) AIP Advances 8:056404
12. Arisawa H, Daimon S, Oikawa Y, Seo Y-J, Harii K, Oyanagi K, Saitoh E (2019) Appl Phys Lett 114:122402

Chapter 3
Magnetomechanical Coupling in Magnetic Micro Mechanical Oscillator

A mechanical oscillator such as a cantilever is widely applied to a sensitive mechanical sensor because of its high Q factor [1]. Moreover, recent advances in microfabrication technologies have made it possible to fabricate mechanical oscillators made of magnetic materials [2, 3], which enables us to directly detect the magnetomechanical coupling via the spin-lattice interaction in the oscillator. In this study, we investigated the coupling between magnetization and mechanical vibration in a micro mechanical oscillator, which is made of a ferrimagnetic insulator yttrium iron garnet $Y_3Fe_5O_{12}$ (YIG) [4]. The mechanical experiment clarified that the mechanical vibration of the YIG oscillator changes depending on its magnetization process. Our theoretical calculation based on the magnetoelastic energy revealed that the observed mechanical-vibration modulation is caused by the magnetization rotation in the YIG oscillator. In this chapter, we describe the details of the experiment and the theoretical analysis.

3.1 Fabrication of YIG Micro Bridge by FIB Method

In this subsection, we explain the fabrication process of a YIG micro bridge structure by using the focused ion beam (FIB) method.

3.1.1 Fabrication Process of YIG Micro Bridge

The prepared sample is a commercially obtained single crystalline YIG (112 μm in thickness) deposited on paramagnetic gadolinium gallium garnet. To prevent possible sample charging caused by the Ga^+ beam irradiation, we deposited a gold film with the thickness of 5 nm on the YIG surface. We fixed the YIG sample to the sample stage in the FIB device by connecting the sample surface to the stage with a conductive tape,

H. Arisawa, *A Study on Spin Mechanics Phenomena in Spin-Lattice Coupled Systems*, Springer Theses, https://doi.org/10.1007/978-981-97-8642-8_3

and performed the microfabrication of the YIG micro bridge. Figure 3.1 shows the fabrication process. First, by irradiating an incident Ga^+ beam to the sample surface, we fabricated a rectangular wall structure in the YIG substrate [Fig. 3.1a]. Next, we irradiated a Ga^+ beam to the wall structure from the angle of 38°, and removed the lower part of the structure to fabricate it into the bridge structure [Fig. 3.1b]. Finally, by irradiating an Ar beam to the sample surface as surface treatment, we removed the YIG layer with the thickness of 50 nm [Fig. 3.1c]. In Fig. 3.1d, we show a scanning electron microscope image of the YIG micro bridge after all the processes, which has the high aspect ratio (length $l = 70\ \mu m \gg$ width $w = 5.7\ \mu m$). Figure 3.1e shows the cross-sectional area of the bridge expected from the fabrication process. Since the minimum incident angle of the Ga^+ beam is 38° due to the limitations of the FIB device, the cross-section is a pentagon. For simplicity, in the following, we assume the sample shape to be a rectangle with the width w and the thickness t_{eff} [Fig. 3.1f], where $t_{eff} = (t_{tot} + t_s)/2$ is the average of the maximum and minimum thickness t_{tot} and t_s of the cross-sectional area, respectively.

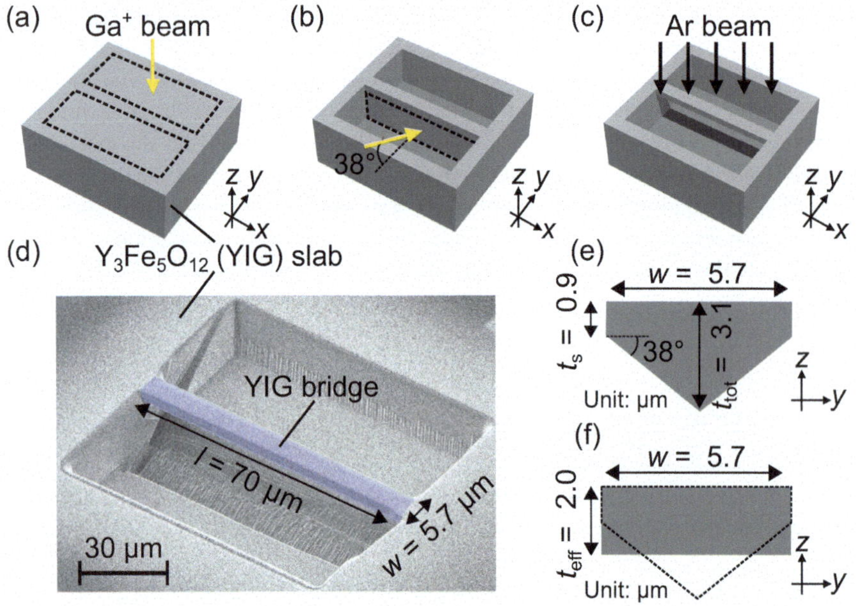

Fig. 3.1 **a–c** Schematic illustrations of the fabrication process of the YIG micro bridge. **a** Ga^+ milling for the two regions surrounded by the dashed lines, **b** Ga^+ milling for the understructure of the center bar, and **c** Ar milling for surface treatment. **d** A scanning electron microscopy image of the YIG micro bridge. l and w are the length and the width of the YIG micro bridge, respectively. **e** A schematic illustration of the cross-sectional area of the YIG micro bridge along the y-z plane. t_s is the side thickness, and t_{tot} is the thickness from the top to the bottom of the pentagonal cross-section. **f** A schematic illustration of the rectangular cross-section with the effective thickness $t_{eff} = (t_{tot} + t_s)/2$. Reproduced from Appl. Phys. Lett. **114**, 122402, (2019) [4]; copyright 2019 AIP Publishing

3.1.2 Evaluation of Sample Damage by Ga^+ Ion Beam Irradiation

In the fabrication process, we carried out the surface treatment with the Ar milling. In this subsection, we explain the main effect of the surface treatment by discussing the YIG magnetization process. The main aim of the Ar milling is to remove the influence of the possible Ga^+ contamination into the sample surface, which can change the magnetic properties of the sample [5]. To check the influence, we evaluated the magnetic property of the YIG substrate by comparing its magnetization process before and after the Ga^+ beam irradiation. The upper panels in Fig. 3.2 show the experimental setup. Here, we estimated the in-plane magnetization process of the YIG by measuring the magnetic Kerr effect. The lower panel in Fig. 3.2a shows the in-plane magnetic field H dependence of the magnetic Kerr rotation angle for the YIG sample without FIB microfabrication. A steep change in the Kerr rotation angle at $H \approx 0$ is observed and the Kerr rotation angle remains almost unchanged above 30 Oe, reflecting the in-plane magnetization process of the YIG. From the result, we evaluated that the saturation magnetic field H_{s0} of the YIG substrate is about 30 Oe when the in-plane magnetic field is applied. Figure 3.2b shows the H dependence of the Kerr rotation angle after microfabrication (1 μm in depth) by the Ga^+ beam irradiation. After the irradiation, the sign of the Kerr rotation angle reverses and the rotation angle does not saturate even if the magnetic field value exceeds H_{s0}. Since the rotation angle is measured without changing the laser position before and after the

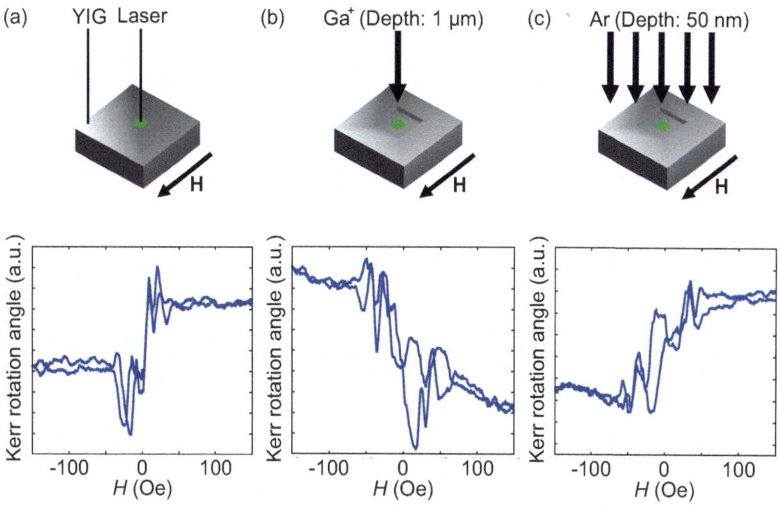

Fig. 3.2 **a–c** The in-plane magnetic field H dependence of the magnetic Kerr rotation angle of the YIG substrate. **a** Before the Ga^+ beam irradiation, **b** After the Ga^+ beam irradiation (1 μm in depth), **c** After the Ar beam irradiation (50 nm in depth). The laser position was fixed before and after the Ga^+/Ar beam irradiation

irradiation, the obtained result cannot be explained by the spatial dependence of the magnetization process and implies that the magnetization process is modulated by the Ga^+ beam irradiation. Figure 3.2c shows the experimental results after removing the fabricated position by 50 nm by the Ar milling. We can see that the sign of the Kerr rotation angle returned to the original one and the saturation magnetic field is almost the same as that before the Ga^+ beam irradiation. From the result, we concluded that the spatial range of the possible magnetic property modulation by the Ga^+ beam irradiation is in the order of several tens of nm, which can be removed by the Ar milling.

3.2 Mechanical Property of YIG Micro Bridge

3.2.1 Mechanical Vibration Spectrum

To check the basic mechanical property of the fabricated YIG micro bridge, we measured mechanical vibration spectra at a zero magnetic field. Figure 3.3a shows the measurement setup. To induce the mechanical vibration of the YIG micro bridge, we placed the bridge on a piezoelectric device and applied an a.c. voltage with the frequency f_P to the piezoelectric device as an input. In response to the a.c. voltage, the piezoelectric device mechanically vibrates in out-of-plane direction at the frequency f_P via the piezoelectric effect. We measured the out-of-plane mechanical vibration spectra of the YIG micro bridge by using a laser Doppler vibrometer. Here, the laser position was fixed at the center of the bridge except for the measurement of the laser position dependence of the mechanical vibration. Figure 3.3b shows the vibration amplitude A spectrum of the YIG micro bridge when the frequency f_P of the piezoelectric device was swept in the range of 2.535 MHz to 2.575 MHz. A clear peak structure was observed around $f_P = 2.555$ MHz, which is well fitted to the Lorentzian function (solid red line), indicating that the YIG micro bridge is in mechanical resonance. From the fitting value of the peak position, we confirmed that the resonance frequency f_1 of the mechanical vibration is 2.555 MHz.

3.2.2 Spatial Profile of Mechanical Vibration

Figure 3.3c shows the position x dependence of the vibration amplitude A, measured by moving the laser spot of the laser Doppler vibrometer along the length of the bridge. Note that the input piezoelectric frequency is fixed at $f_P = f_1$ to excite the resonant mechanical vibration in the bridge. The value of A takes its maximum at the center of the bridge and decreases towards both ends, which means that the observed mechanical vibration is the first-order resonance mode. By fitting the result to the theoretical formula (1.42) of the first resonance mode, derived in Chap. 1, the

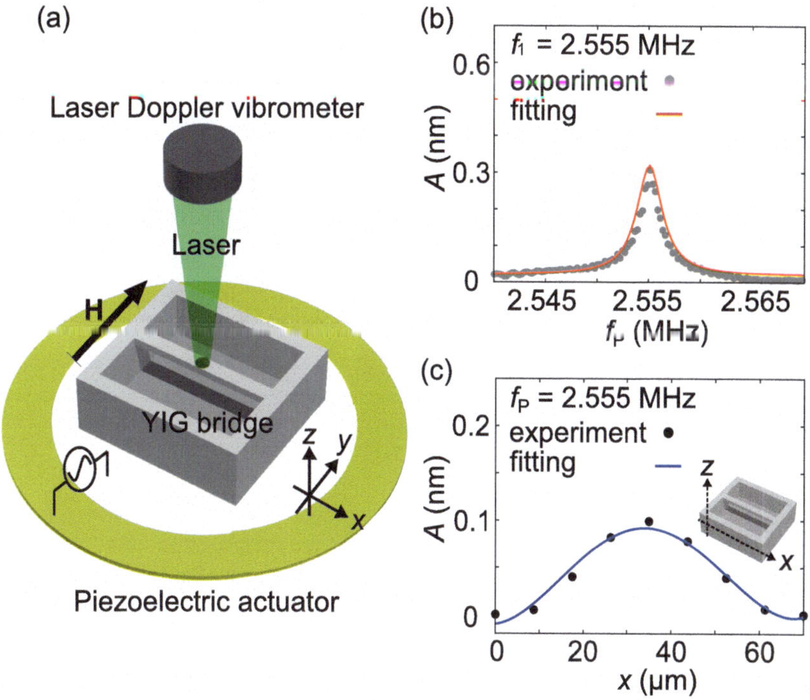

Fig. 3.3 **a** A schematic illustration of the measurement setup. **H** denotes the external in-plane magnetic field. **b** The piezoelectric actuator frequency f_P dependence of the out-of-plane vibration amplitude A of the YIG micro bridge (gray dots). A Lorentzian fitting curve is shown in red color. f_1 is the resonance frequency obtained from the fitting. **c** Spatial distribution of A (black dots). A fitting curve is shown in blue color. f_P was set to 2.555 MHz in the measurement. Adapted from Appl. Phys. Lett. **114**, 122402, (2019) [4]; copyright 2019 AIP Publishing

wavelength λ_1 of this mode is determined as $\lambda_1 = 93$ μm. By substituting $\lambda_1 = 93$ μm and each parameter of the YIG micro bridge into the formula (1.46), we can obtain the theoretical value $f_{th} = 2.6$ MHz as the first mode resonance frequency. This value agrees well with the experimental value $f_1 = 2.555$ MHz, supporting our interpretation that the first resonance mode of the mechanical vibration with the frequency f_1 is excited in the YIG micro bridge.

3.3 Magnetic Property of YIG Micro Bridge

To evaluate the magnetic properties of the YIG micro bridge, we measured the magnetization process of the bridge by using the magneto-optical microscope. Here, the laser position was fixed at the center of the YIG micro bridge. Figure 3.4 shows the magnetic field H dependence of the Kerr rotation angle, where H was applied

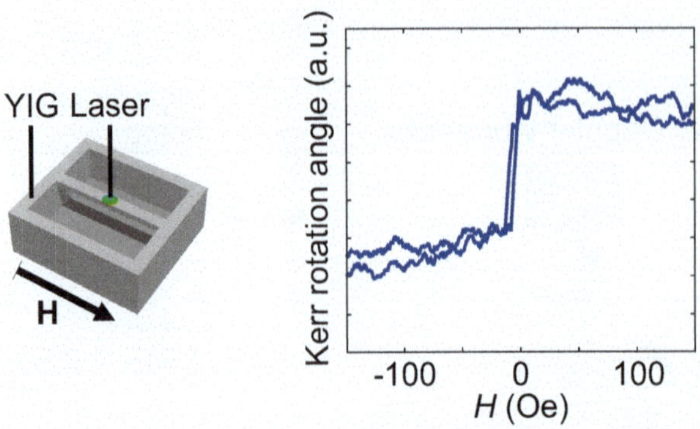

Fig. 3.4 The magnetic field dependence of the magnetic Kerr rotation angle of the YIG micro bridge. The magnetic field **H** was applied in the sample plane and in the longitudinal direction of the bridge, and the laser spot was focused at the center of the bridge

along with the length of the bridge. Similar to the magnetization curve before the microfabrication, the Kerr rotation angle is almost constant after exhibiting a sharp step structure near the zero magnetic field. The value of the saturation magnetic field H_{sx} is also similar to the value before microfabrication, implying that only a small demagnetizing field due to the shape magnetic anisotropy acts in the longitudinal direction of the bridge. To quantitatively evaluate the shape anisotropy, we calculate the demagnetization field from the shape of the bridge. Here, we approximately regarded the YIG micro bridge as an ellipsoidal structure with the three main axes of $l = 70\,\mu\text{m}$ (bridge length), $w = 5.7\,\mu\text{m}$ (bridge width), and $t_{\text{eff}} = 2.0\,\mu\text{m}$ (effective bridge depth), and calculated the demagnetization coefficients N_x, N_y, and N_z in the x, y, and z directions of the bridge, respectively,

$$N_x = 0.01, \quad N_y = 0.26, \quad N_z = 0.73. \tag{3.1}$$

Here, $N_x < N_y$ reflects the in-plane shape magnetic anisotropy, showing that the longitudinal (transverse) direction of the bridge is the easy (hard) axis of its magnetization process. The value of N_x gives the longitudinal demagnetization field $H_{\text{d}x}$ as

$$H_{\text{d}x} = N_x M_\text{s} = 1.43 \times 10^3\ \text{A/m} \approx 20\,\text{Oe} \tag{3.2}$$

where $M_\text{s} = 1.43 \times 10^5$ A/m is the saturation magnetization of YIG [6]. The value of $H_{\text{d}x}$ is in rough agreement with the experimentally observed value of H_{sx}. We also estimated the demagnetization field $H_{\text{d}y}$ in the y direction as

$$H_{\text{d}y} = N_y M_\text{s} = 3.71 \times 10^4\ \text{A/m} \approx 470\ \text{Oe}. \tag{3.3}$$

3.4 Magnetic Field Dependence of Mechanical Vibration Spectrum

To investigate magnetomechanical effects on the bridge, we measured the magnetic field dependence of the mechanical vibration spectrum. Figure 3.5a shows the mechanical vibration spectra at each in-plane magnetic field H, whose direction is parallel to the bridge [x direction in Fig. 3.3a]. Here, the characteristics of the spectra greatly changes by the application of the magnetic field. The change in the peak position is remarkable at low magnetic field $\approx$ 10 Oe, which means that the resonance mechanical vibration frequency in the bridge is modulated at the field. To analyze the resonance frequency f_1 change, we defined

$$\Delta f_1(H) = \frac{f_1(H) - f_1(H_{\rm min})}{f_1(H_{\rm min})} \quad (H_{\rm min} < 0 : \text{minimum applied field}) \tag{3.4}$$

and obtained its detailed H dependence [Fig. 3.5b], where the resonance frequency increases abruptly at $\approx$ 10 Oe. Another characteristic is that the resonance frequency takes a constant saturation value at a high magnetic field. The magnetic field reaching the saturation value is comparable to the measured saturation field of the YIG micro bridge, suggesting the correlation between the resonance frequency and the magnetization process of the bridge.

Figure 3.6 shows the H dependence of Δf_1 when H is applied in the y direction. As is different from the case of $\mathbf{H} \parallel x$, Δf_1 shows non-monotonic changes depending on H. f_1 increases with the application of the field from $H = 0$, then decreases,

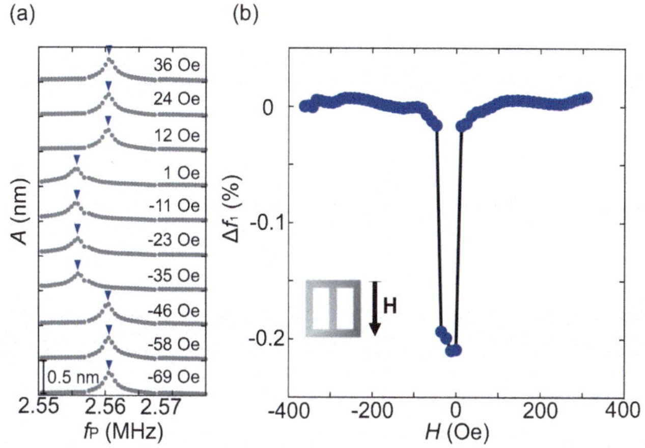

Fig. 3.5 **a** The piezoelectric actuator frequency $f_{\rm P}$ dependence of the vibration amplitude A at each magnetic field. The magnetic field **H** was applied in the sample plane and in the longitudinal direction of the bridge. **b** The magnetic field strength H dependence of the resonance frequency shift Δf_1 (blue dots)

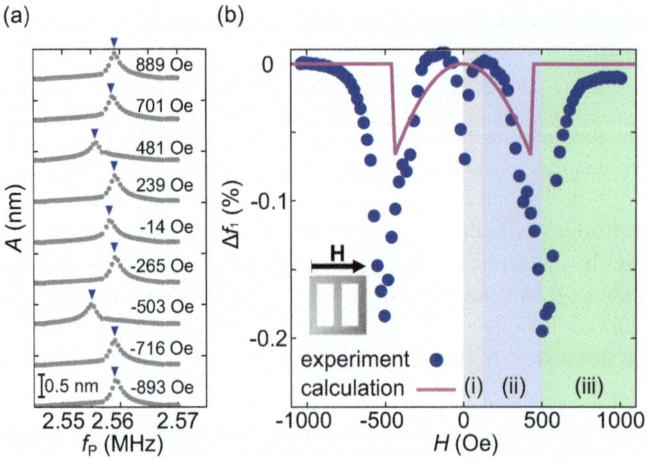

Fig. 3.6 **a** The piezoelectric actuator frequency f_P dependence of the vibration amplitude A at each magnetic field. The magnetic field **H** was applied in the sample plane and in the perpendicular direction to the bridge. **b** The magnetic field strength H dependence of the resonance frequency shift Δf_1 (blue dots). A purple solid curve is a result of theoretical model calculation

and increases again around $H = 500$ Oe. When $H > 500$ Oe, the value of Δf_1 reaches to the saturation value, which is the same as the case of $\mathbf{H} \parallel x$. To check the correspondence between the frequency change and the magnetization process, we estimate the saturation field H_{sy} in the y direction from the demagnetization field H_{dy} calculated in the previous section, $H_{sy} \approx H_{dy} = 470$ Oe. As in the case of the $\mathbf{H} \parallel x$, the value of $H_{sy} = 470$ Oe shows good agreement with the magnetic field value $H \approx 500$ Oe at which f_1 begins to saturate. These results indicate that the mechanical resonance frequency of the YIG micro bridge changes according to the magnetization process of the bridge.

3.5 Theoretical Analysis

Magnetoelastic coupling is one of possible origins of the observed resonance-frequency modulation depending on the magnetization process. In this section, we analyze the observed signal by considering a phenomenological model based on the magnetoelastic coupling in the YIG micro bridge.

3.5.1 Delta-E Effect: Change in Young's Modulus Due to Magnetization Rotation

In the presence of the magnetoelastic coupling, elastic properties change associated with the magnetization in a magnet. One typical example is the Young's modulus, a parameter that represents the hardness of an elastic body. Due to the magnetostriction in a magnet, the Young's modulus is known to depend on the magnetization process, which is called the delta-E effect [7, 8]. The change in the Young's modulus causes the modulation of the mechanical resonance frequency. In the present case of the YIG micro bridge, the relation between the Young's modulus and the mechanical resonance frequency is given by

$$f_1 = \frac{2\pi t_{\text{eff}}}{\lambda_1^2}\sqrt{\frac{E}{12\rho}}, \tag{3.5}$$

In the following, we consider the possible origin of the observed results based on the magnetization process and the delta-E effect.

First we discuss the experimental results at $\mathbf{H} \parallel x$. In this configuration, $\mathbf{H}$ is along the easy axis of the magnetic shape anisotropy, and the magnetization is expected to saturate at $H = H_{sx}$ via the magnetic domain disappearance. The delta-E effect due to the change in the domain distribution has been reported in ferromagnetic metals such as Fe and Ni, where the domain disappearance increases the Young's modulus [9], resulting in the increase in the mechanical resonance frequency. The increase in f_1 at $H \approx H_{sx}$ [Fig. 3.5b] is consistent with the previous reports, and the delta-E effect due to the domain disappearance can be attributed to the observed signal.

Next, we consider the case of $\mathbf{H} \parallel y$. The increase in f_1 in the low field region $0 < H < 50$ Oe [region (i) in Fig. 3.6b] can be explained based on the same discussion as that for $\mathbf{H} \parallel x$. Here, for the small H, the effective magnetic field due to the strong shape magnetic anisotropy is expected to induce the magnetization saturation along the bridge length even when $\mathbf{H} \parallel y$, where the delta-E effect due to the domain disappearance is attributable to the increase in f_1. Note that in the above discussion we assume that the x component of H exists due to the inevitable slight deviation of the field angle from the ideal configuration of $\mathbf{H} \parallel y$. In contrast, the decrease and increase of f_1 in the high field region 50 Oe $< H <$ 1000 Oe [regions (ii) and (iii) in Fig. 3.6b, respectively] cannot be explained by the domain disappearance. In the high field region, the magnetization rotation from the easy axis to the hard axis should occur due to the competition between the shape magnetic anisotropy and the external magnetic field, indicating a delta-E effect governed by the shape magnetic anisotropy. In the following, we focus on the measurement results for $\mathbf{H} \parallel y$ in a high magnetic field region [region (ii) and (iii) in Fig. 3.6b].

3.5.2 Phenomenological Model for Delta-E Effect

To verify the resonance frequency modulation effect by the magnetization rotation, we carried out theoretical analysis using the phenomenological model based on the delta-E effect [10]. A schematic illustration of the model is shown in Fig. 3.7. **H** and θ are the external magnetic field along the y-axis and the angle between the magnetization **M** and the x-axis, respectively. $\boldsymbol{\sigma}$ denotes the stress along the bridge length, which originates from the elastic deformation of the bridge due to the mechanical vibration. For simplicity, we assume the magnetization rotation in a single magnetic domain.

We define the free energy F of the YIG bridge in the model by considering the Zeeman energy F_{Zeeman}, the shape magnetic anisotropy energy F_{ani}, and the magnetoelastic energy F_{me} as

$$F = F_{\mathrm{Zeeman}} + F_{\mathrm{ani}} + F_{\mathrm{me}} \tag{3.6}$$

$$F_{\mathrm{Zeeman}} = -\mu_0 M_{\mathrm{s}} H \sin\theta \tag{3.7}$$

$$F_{\mathrm{ani}} = K \sin^2\theta \tag{3.8}$$

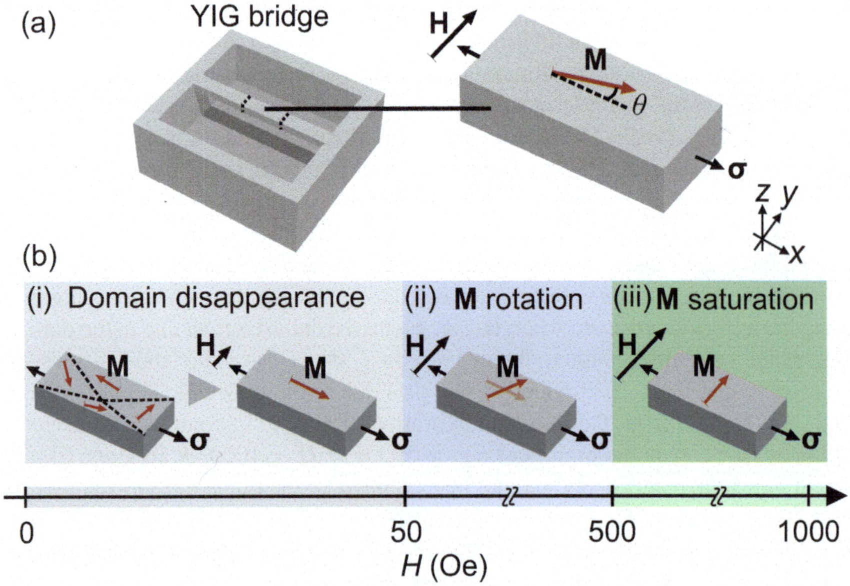

Fig. 3.7 **a** A schematic illustration of the calculation model for the delta-E effect based on the magnetization rotation. **H**, **M**, σ, and θ are the magnetic field, the magnetization, the tensile stress along the bridge, and the relative angle between **M** and the longitudinal axis of the bridge, respectively. **b** A schematic illustration of the magnetization process of the bridge in each H region

$$F_{\text{me}} = -\frac{3}{2}\lambda_s\sigma\left(\cos^2\theta - \frac{1}{3}\right), \tag{3.9}$$

where $\mu_0 = 4\pi \times 10^{-7}$ N/A^2 and $\lambda_s = -2.2 \times 10^{-6}$ [11] are the permeability in vacuum and the saturation magnetostriction of YIG, respectively. K is defined as $\mu_0(N_y - N_x)M_s{}^2/2$.

The Young's modulus E of a rigid body is given by

$$\frac{1}{E} = \frac{\partial\varepsilon}{\partial\sigma}, \tag{3.10}$$

where ε denotes elastic strain. Here, in the presence of magnetostriction ε_{ms}, the Young's modulus E in Eq. (3.10) depends on the magnetization angle θ as

$$\frac{1}{E} = \frac{\partial\varepsilon}{\partial\sigma} = \frac{\partial(\varepsilon_{\text{el}} + \varepsilon_{\text{ms}}(\theta))}{\partial\sigma} = \frac{\partial\varepsilon_{\text{el}}}{\partial\sigma} + \frac{\partial\varepsilon_{\text{ms}}(\theta)}{\partial\sigma} = \frac{1}{E_{\text{el}}} + \frac{\partial\varepsilon_{\text{ms}}(\theta)}{\partial\sigma}, \tag{3.11}$$

where ε_{el} and E_{el} are the strain and the Young's modulus without the magnetoelastic coupling, respectively. From the above equation, we can derive the change in the Young's modulus as

$$\frac{E}{E_{\text{el}}} = \frac{1}{1 + E_{\text{el}}\frac{\partial\varepsilon_{\text{ms}}(\theta)}{\partial\sigma}}. \tag{3.12}$$

Now we are ready to calculate the magnetic field dependence of the resonance frequency of the YIG bridge based on the model. As a first step, we minimize F to obtain $\theta(H, \sigma)$ as

$$\frac{\partial F}{\partial\theta} = \{(2K + 3\lambda_s\sigma)\sin\theta - \mu_0 M_s H)\}\cos\theta = 0, \tag{3.13}$$

which leads to

$$\sin[\theta(H, \sigma)] = \frac{\mu_0 M_s H}{2K + 3\lambda_s\sigma} \qquad (0 < |H| < H_1) \tag{3.14}$$

$$\theta(H, \sigma) = 90^\circ, \qquad (H_1 < |H|) \tag{3.15}$$

where $H_1 = \frac{2K+3\lambda_s\sigma}{\mu_0 M_s}$. From the magnetization rotation angle $\theta(H, \sigma)$ obtained above and the magnetostriction formula (1.72), we can derive

$$\frac{\partial\varepsilon_{\text{ms}}(\theta)}{\partial\sigma} = \frac{(3\lambda_s\sin\theta)^2}{2K + 3\lambda_s\sigma} \tag{3.16}$$

Substitution of Eqs. (3.14), (3.15), and (3.16) to Eq. (3.12) gives us E as a function of (H, σ):

$$\frac{E}{E_{\mathrm{el}}} = \begin{cases} \frac{1}{1+\frac{E_{\mathrm{el}}(3\mu_0\lambda_S M_S H)^2}{(2K+3\lambda_S\sigma)^3}} & (0 < |H| < H_1) \\ 1 & (H_1 < |H|) \end{cases}$$

By using Eq. (3.5), we define the resonance frequency change Δf_{cal} due to the E change as

$$\Delta f_{\mathrm{cal}} = \frac{f_1 - f_{\mathrm{el}}}{f_{\mathrm{el}}} = \frac{f_1}{f_{\mathrm{el}}} - 1 = \sqrt{\frac{E}{E_{\mathrm{el}}}} - 1. \tag{3.17}$$

Here, f_{el} denotes the mechanical resonance frequency without the magnetoelastic coupling. By substituting E/E_{el} to the above equation, we can derive the final form of Δf_{cal}:

$$\Delta f_{\mathrm{cal}} = \begin{cases} \sqrt{\frac{1}{1+\frac{E_{\mathrm{el}}(3\mu_0\lambda_S M_S H)^2}{(2K+3\lambda_S\sigma)^3}}} - 1 & (0 < |H| < H_1) \\ 0 & (H_1 < |H|) \end{cases}$$

Figure 3.6b shows the calculation result of Δf_{cal} as a function of the magnetic field (purple solid curve). Compared to the experimental results (blue dots), the decrease (increase) in f_1 before (after) $H \approx 500$ Oe are reproduced well in the calculation, supporting our interpretation that the observed signal originates from the Young's modulus modulation mediated by the magnetization rotation.

3.5.3 *Physical Origin*

Now, we discuss the mechanism of the observed signal by comparing between the experiment and the phenomenological model. We first consider the magnetic field range 50 Oe $< H <$ 500 Oe at which f_1 decreases [region (ii) in Figs. 3.6b and 3.7b]. This range corresponds to the magnetization rotation region ($0° < \theta < 90°$) before the magnetization saturation. In general, when stress $\boldsymbol{\sigma}$ is applied to a magnet, the magnetization rotates via the magnetoelastic coupling to reduce the free energy F. In the case of the YIG micro bridge with the negative λ_s value, when $\boldsymbol{\sigma}$ is applied as shown in Fig. 3.7a, θ increases to minimize F. Since the increase in θ induces extra strain via the magnetostriction in addition to the elastic tensile strain, the Young's modulus E of the YIG micro bridge decreases. Considering that the increase in θ is due to the competition between $F_{\mathrm{Zeeman}} + F_{\mathrm{me}}$ (most stable at $\theta = 90°$) and F_{ani} (most stable at $\theta = 0°$), we can also see that the θ increase becomes prominent when F_{Zeeman} is more dominant with the increase in H. Therefore, the extra strain becomes greater with increasing H, resulting in the monotonic decrease in E [region (ii) in Fig. 3.6b]. In contrast, after the magnetization is saturated perpendicular to the bridge at the region of 500 Oe $< H$ [region (iii) in Figs. 3.6b and 3.7b], the value of θ remains constant ($= 90°$) regardless of the magnetic field. Magnetostriction does not occur unless the magnetization angle changes, and the contribution of magnetostriction disappears abruptly right after the magnetization saturation, leading to no modulation of the Young's modulus. Since the Young's modulus returns to its original value from the decreased value, the resonance frequency shows a sharp increase at the saturation magnetic field.

3.6 Summary

In this study, we explored the correlation between the magnetization process and the mechanical vibration in a YIG micro bridge. By using the focused ion beam method, we fabricated a magnetic insulator YIG into a micro bridge structure, which mechanically vibrates in resonance. With laser Doppler vibrometry, we found that the mechanical resonance frequency of the YIG micro bridge changes according to the magnetization process in the bridge. By comparing the obtained result with model calculation, we verified that the resonance frequency change originates from the Young's modulus modulation mediated by the magnetization rotation in the bridge via the magnetoelastic coupling. YIG is widely used in spintronics due to its small magnetic damping [12, 13], and the YIG oscillator demonstrated here could be a platform to access magnetomechanical coupling between spin waves and mechanical vibrations.

References

1. Nguyen CTC, Howe RT (1999) IEEE J Solid State Circuits 34:440
2. Xiong G (2001) Appl Phys Lett 79:3461
3. Oikawa Y, Arisawa H, Daimon S, Saitoh E (2018) Appl Phys Lett 113:142407
4. Arisawa H, Daimon S, Oikawa Y, Seo Y-J, Harii K, Oyanagi K, Saitoh E (2019) Appl Phys Lett 114:122402
5. Park C-M, Bain JA (2002) IEEE Trans Mag 38:2237
6. Hansen P, R'oschmann P, Tolksdorf W (1974) J Appl Phys 45:2728
7. Chikazumi s (ed) (2009) Physics of ferromagnetism. Oxford Science Publications, Oxford
8. Gojdka B, Jahns R, Meurisch K, Greve H, Adelung R, Quandt E, R. Knöchel, and F. Faupel (2011) Appl Phys Lett 99, 223502
9. Morales AL, Nieto AJ, Chicharro JM, Pintado P (2008) Meas Sci Technol 19:125702
10. Squire PT (1990) J Magn Magn Mater 87:299
11. Clark AE, DeSavage B, Coleman W, Callen ER, Callen HB (1963) J Appl Phys 34:1296
12. Stancil DD, Prabhakar A (ed) (2009) Spin waves. Springer
13. Arisawa H, Daimon S, Oikawa Y, Kikkawa T, Saitoh E (2023) Phys Rev B 107:134408

Chapter 4
Spin Current Volume Effect in Spin-lattice Coupled System

The magneto-volume effect (MVE), the volume change associated with spin fluctuation, has been a central issue in the physics of magnetism for a long time [1, 2]. Since the discovery of the MVE, extensive studies on the MVE have made remarkable progress in the physics of spin fluctuation and electronic correlation in ferromagnets. By contrast, recent spintronics studies have revealed that a spin current, a flow of spin angular momentum, is a powerful tool for controlling spin fluctuation [3], which should enable us to manipulate the volume of a magnet by using a spin current. In this study, we tried to demonstrate the volume modulation effect in a magnetostrictive material mediated by a spin current, a spin current volume effect [4]. By injecting a spin current into the ferromagnetic $Tb_{0.3}Dy_{0.7}Fe_2$ film, we investigated its volume change in response to the spin current injection. Theoretical analysis using the fluctuation dissipation theorem demonstrated that the observed spin-current volume change is caused by the modulation of fluctuation in the spin system. In this chapter, we explain the details of the experimental results and the theoretical calculation.

4.1 Spin Current Volume Effect

Figure 4.1 shows a schematic illustration of the concept of the spin-current-induced volume modulation effect. At finite temperature, the magnetization in a ferromagnet thermally fluctuates around a magnetic field. When a spin current is injected into the ferromagnet, spin angular momentum is transferred from the spin current to the magnetization. As a result of the spin transfer, the spin transfer torque modulates the magnetization fluctuation [3]. In the presence of the magnetoelastic coupling in the magnet, its volume expansion/contraction is expected to appear due to the magnetization fluctuation change.

H. Arisawa, *A Study on Spin Mechanics Phenomena in Spin-Lattice Coupled Systems*, Springer Theses, https://doi.org/10.1007/978-981-97-8642-8_4

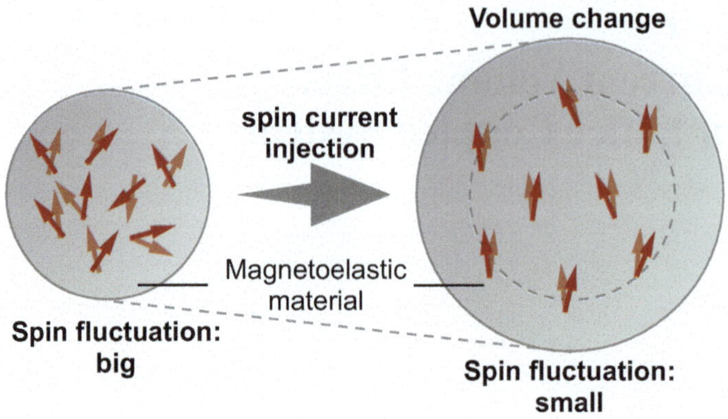

Fig. 4.1 A schematic illustration of a spin current volume effect. The volume of a magnetoelastic material changes due to the spin fluctuation change via the spin current injection

To explore the spin current volume effect, we focused on a giant magnetostrictive material $Tb_{0.3}Dy_{0.7}Fe_2$ [5]. The ferromagnet $Tb_{0.3}Dy_{0.7}Fe_2$ is known to exhibit the structural phase transition accompanied by the magnetic anisotropy transition near room temperature [6], resulting in its large magnetostriction. Therefore, $Tb_{0.3}Dy_{0.7}Fe_2$ can be a good playground to investigate the magnetomechanical response to the spin current injection.

Figure 4.2 shows a schematic illustration of the measurement setup in the present study. We prepared a $Tb_{0.3}Dy_{0.7}Fe_2$/(Pt, Cu, or W) bilayer system. The in-plane magnetic field $\mathbf{H}$ was applied perpendicular to the sample length (z direction) except for the measurement of the magnetic field angle dependence. As an input, we applied an a.c. current $\mathbf{I}_{AC}$ (the intensity I_{AC}) in the paramagnetic metal layer (Pt, Cu, or W) along its length (x direction). Here, when an electric current flows in the $+x$ direction in the paramagnetic metal, the spin Hall effect induces a spin accumulation $\boldsymbol{\sigma}$ in the $-z$ direction at the bilayer interface [left panel in Fig. 4.2a]. This spin accumulation transfers angular momentum to the magnetization $\mathbf{M}$ in the $Tb_{0.3}Dy_{0.7}Fe_2$ film and modulates the magnetization fluctuation. In response to the a.c. current application, the direction of the injected spin current and the spin accumulation changes alternatively with the frequency f_{AC}, leading to the alternative modulation of the magnetization fluctuation and resultant volume changes at f_{AC} [Fig. 4.2b]. As an output of the a.c. current, the out-of-plane (y direction) mechanical vibration spectrum at the $Tb_{0.3}Dy_{0.7}Fe_2$ surface was measured with a laser Doppler vibrometer. The value of f_{AC} was fixed at 10 kHz. All the measurements were performed at room temperature and in a high vacuum $\approx 10^{-4}$ Pa.

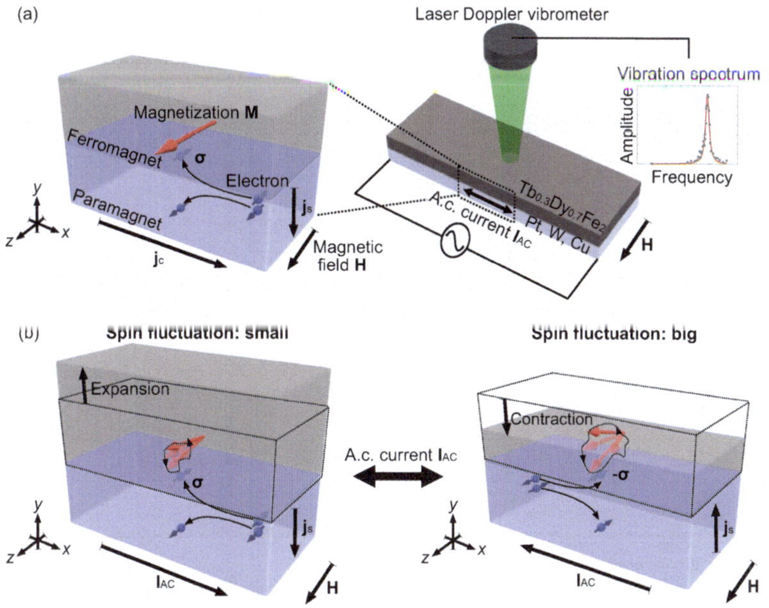

Fig. 4.2 **a** Schematic illustrations of the experimental setup. **M**, σ, $\mathbf{j}_\mathrm{c}$, $\mathbf{j}_\mathrm{s}$, and **H** are the magnetization, the spin polarization vector of a conduction electron, an electric current, a spin current, and the external magnetic field, respectively. An a.c. spin current was injected into the $Tb_{0.3}Dy_{0.7}Fe_2$ film by applying an a.c. current $\mathbf{I}_{AC}$ to the (Pt, W, or Cu) layer, and the mechanical response of the $Tb_{0.3}Dy_{0.7}Fe_2$ film was measured with a laser Doppler vibrometer. **b** Schematic illustrations of the spin fluctuation modulation and the volume change in a ferromagnet when the a.c. current is applied to the paramagnetic metal

4.2 Evaluation of Magnetostrictive Film Properties

Here we measured the magnetization process, the magnetostriction, the resistance, and the composition of the fabricated magnetic film on paramagnetic metals to evaluate its physical properties, and confirmed that the deposited film exhibits characteristics typical for $Tb_{0.3}Dy_{0.7}Fe_2$.

4.2.1 Chemical Composition

We first evaluated the chemical composition of the magnetic films, which were fabricated by the electroplating method [7]. As shown in Figs. 4.3a–c, we carried out X-ray photoelectron spectroscopy (XPS) for the magnetic film deposited on the Pt film and

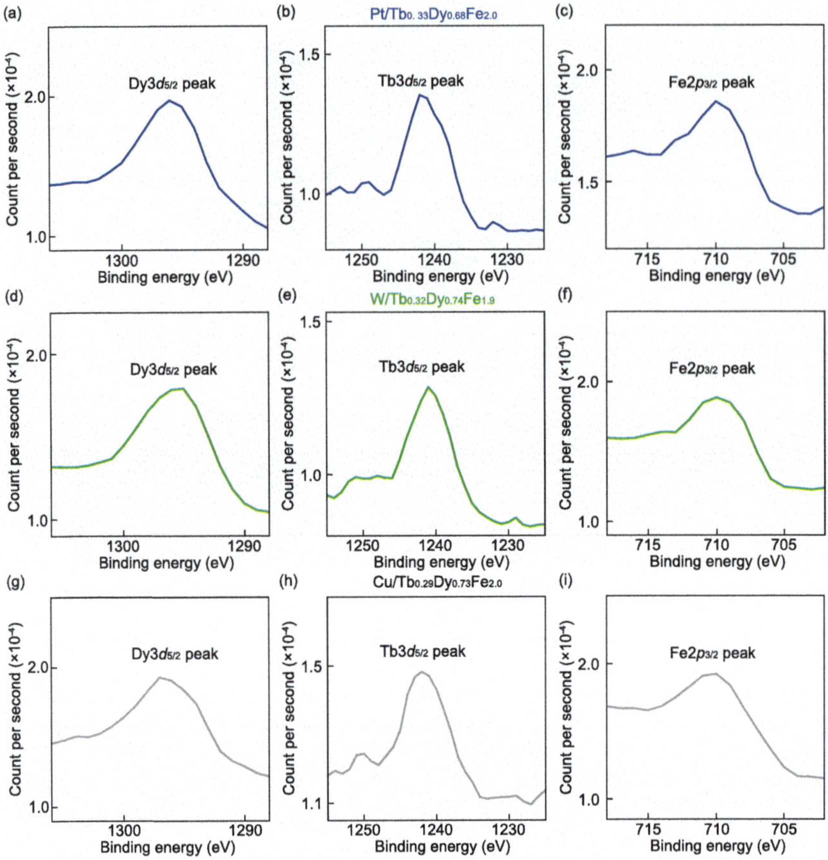

Fig. 4.3 **a–i** XPS spectra for the magnetic films on the **a–c** Pt, **d–f** W, and **g–i** Cu films. Reproduced from Nat. Commun. **13**, 2440 (2022) [4]; copyright 2022 Author(s), licensed under a Creative Commons Attribution (CC BY) license

estimated the atomic composition ratio of the film as Tb:Dy:Fe $\approx$ 11.1 : 22.5 : 66.4. From the result, the chemical composition of the film is given by $\approx Tb_{0.33}Dy_{0.68}Fe_{2.0}$, which is close to the optimal composition for large magnetostriction, $Tb_{0.3}Dy_{0.7}Fe_2$. By measuring the XPS spectra for the magnetic film deposited on the W and Cu films [Figs. 4.3d–f and g–i, respectively], we confirmed the similar composition $\approx$ W/$Tb_{0.32}Dy_{0.74}Fe_{1.9}$ and Cu/$Tb_{0.29}Dy_{0.73}Fe_{2.0}$.

4.2.2 Magnetization Process

We next measured the magnetization curve of the magnetic film. We used a vibrating sample magnetometer in Physical Property Measurement System (PPMS) for the experiment. Figure 4.4a shows the M-H curve for the magnetic film on the Pt film. The magnetic field H was applied in the sample plane. From the result, we estimated that the saturation magnetic field and the coercive field of this sample are 7700 Oe and 240 Oe, respectively, which are comparable to the previous studies [8, 9]. Figure 4.4b shows the M-T curve. Here, the slope of the magnetization curve changes, and the value of M takes a local maximum around $T_{sp} \approx 250$ K. This is a characteristic behavior for the magnetic anisotropy transition at the morphotropic phase boundary described in Chap. 2 [Fig. 2.1b]. The value of $T_{sp} \approx 250$ K is also close to the value reported in the previous study [6], consistent with the properties for $Tb_{0.3}Dy_{0.7}Fe_2$.

4.2.3 Magnetostriction

By measuring the magnetostriction, we evaluated the magnetoelastic properties of the magnetic film. Figure 4.5a shows a schematic illustration of the measurement. We first fabricated a magnetic film on Cu into a cantilever structure. When an external magnetic field H is applied in the longitudinal direction of the cantilever, magnetostriction appears in the magnetic film. Since the magnetostriction occurs only in the upper magnetic film, the cantilever is distorted vertically according to the magnetostriction of the film. By measuring the displacement of the cantilever in the y

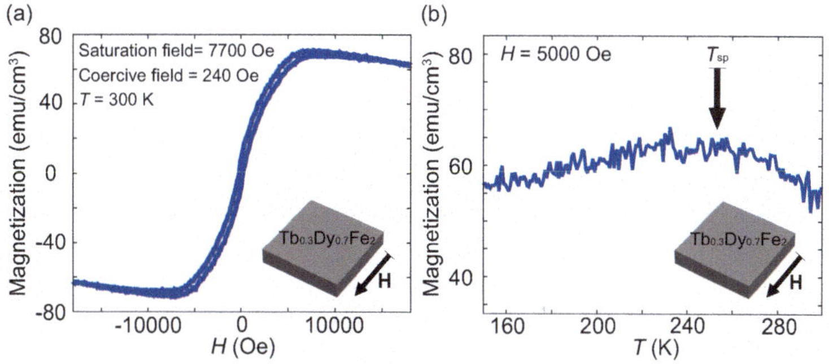

Fig. 4.4 **a** The in-plane magnetic field H dependence of the magnetization of the magnetic film. The saturation magnetic field and the coercive field are 7700 Oe and 240 Oe, respectively. **b** Temperature T dependence of the magnetization of the magnetic film. The value of H was set to 5000 Oe. Around the temperature $T_{sp} \approx 250$ K indicated by the black arrow, the magnetization takes a local maximum value, which is similar to the previous study [6]

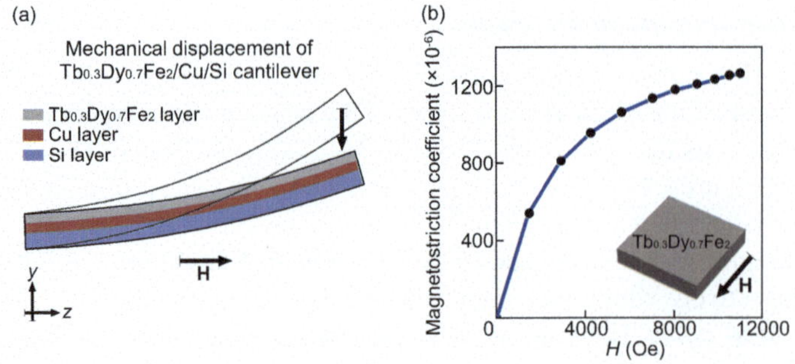

Fig. 4.5 **a** A schematic illustration of the magnetostriction measurement. **b** The magnetic field *H* dependence of the magnetostriction of the magnetic film. The magnetic field **H** is applied in the sample plane. Adapted from Nat. Commun. **13**, 2440 (2022) [4]; copyright 2022 Author(s), licensed under a Creative Commons Attribution (CC BY) license

direction under the magnetic field, the magnetostriction of the film was estimated. The cantilever structure was microfabricated with the support of Prof. Takahito Ono and Dr. Hang Shim (Ph.D. student at the time) in Tohoku University. Figure 4.5b shows the H dependence of the magnetostriction. The value of the magnetostriction increases with the increase in H. The magnetostriction $\approx$1200 ppm near the saturation field is comparable to the saturation magnetostriction $\approx$1700 ppm of the bulk $Tb_{0.3}Dy_{0.7}Fe_2$ [9], indicating that the fabricated film exhibits the large spin-lattice interaction.

4.2.4 *Resistivity*

Here we measured the resistivity of the deposited film. For the efficient spin-current injection into the $Tb_{0.3}Dy_{0.7}Fe_2$ film via the spin Hall effect, the resistance of the $Tb_{0.3}Dy_{0.7}Fe_2$ film should be greater than that of the paramagnetic films, where a large portion of an a.c. current applied in the sample flows in the paramagnetic films and converted to an a.c. spin current. By attaching a short needle of a prober (Toei Kagaku Sangyo Co., Ltd.) in contact with the $Tb_{0.3}Dy_{0.7}Fe_2$ film surface we estimated that the lower limit of the resistivity is in the order of 10^{-4} $\Omega\cdot$m. Compared to the typical resistivity values for Pt, W, and Cu, the resistivity of the $Tb_{0.3}Dy_{0.7}Fe_2$ film is three orders of magnitude higher. In all the samples the ratio between the cross-sectional area and the length of the films are almost the same for the $Tb_{0.3}Dy_{0.7}Fe_2$ and paramagnetic films, and the leakage current flowing through the $Tb_{0.3}Dy_{0.7}Fe_2$ film is estimated to be less than 1/1000 of the input current.

4.3 Spin-Current Mechanical Response in $Tb_{0.3}Dy_{0.7}Fe_2$

The above evaluation of physical properties confirmed that the fabricated magnetic film exhibits typical characteristics as $Tb_{0.3}Dy_{0.7}Fe_2$. In this section, we discuss the mechanical response of the $Tb_{0.3}Dy_{0.7}Fe_2$ film to the spin-current injection.

4.3.1 *Mechanical Vibration Spectrum*

Figure 4.6 shows the frequency spectra of the out-of-plane vibration amplitude and its phase for the $Tb_{0.3}Dy_{0.7}Fe_2$(60 nm)/Pt(140 nm) sample at $H = 1530$ Oe and $I_{AC} = 50$ mA. A vibration peak appears at $f = f_{AC}$ and the phase near the peak is fixed at about 0°. The result indicates that the surface of the $Tb_{0.3}Dy_{0.7}Fe_2$ film mechanically vibrates at the same frequency as that of the a.c. current with the constant phase delay. Note that the observed signal cannot be explained by thermal expansion due to the Joule heating, which is proportional to the square of the current ($f = 2f_{AC}$ term). In the following, we performed several control experiments to clarify the mechanism of the observed mechanical signal.

4.3.2 *Field Dependence of Observed Mechanical Signal*

Figure 4.7a (solid red curves) shows the vibration spectra when the magnetic field H is reversed from $H = +1530$ Oe to $H = -1540$ Oe. For comparison, the results at $H = +1530$ Oe shown in the previous section are also plotted by the blue solid

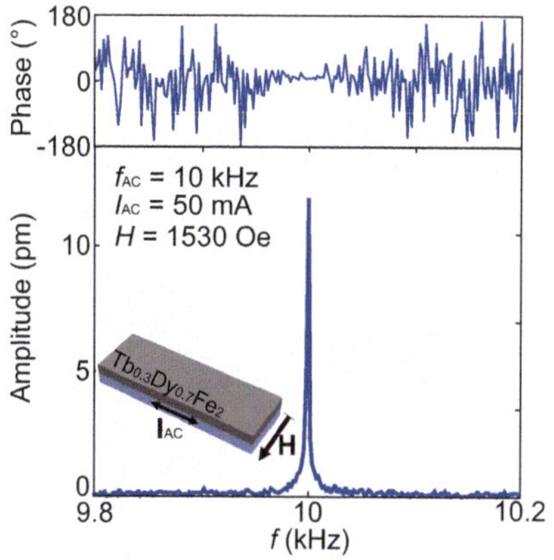

Fig. 4.6 The frequency f spectra of the vibration amplitude and phase for the $Tb_{0.3}Dy_{0.7}Fe_2$/Pt sample. f_{AC}, I_{AC}, and H denote the frequency of the a.c. current, the amplitude of the a.c. current, and the magnetic field, respectively

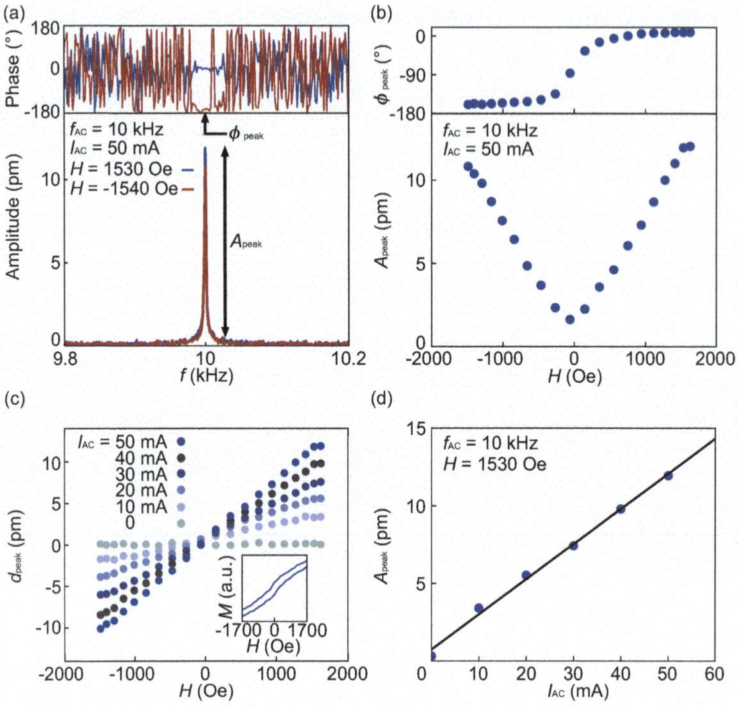

Fig. 4.7 **a** The frequency f spectra of the vibration amplitude and phase for the $Tb_{0.3}Dy_{0.7}Fe_2$/Pt sample when the magnetic field direction is reversed (red solid curve). Compared to the result before the field reversal (blue solid curve), the peak amplitude A_{peak} at $f = f_{AC}$ remains unchanged, while the phase ϕ_{peak} at $f = f_{AC}$ changes by about 180°. **b** The magnetic field H dependence of A_{peak} and ϕ_{peak}. **c** The H dependence of the vibration displacement $d_{peak} = A_{peak}\cos\phi_{peak}$ at each a.c. current amplitude I_{AC}. The inset shows the M-H curve of the $Tb_{0.3}Dy_{0.7}Fe_2$ film. **d** The I_{AC} dependence of A_{peak} at $H = 1530$ Oe

curves. Comparing the results before and after the field reversal, we found that, while the amplitude at $f = f_{AC}$, A_{peak}, remains almost unchanged, the phase at $f = f_{AC}$, ϕ_{peak}, changes by about 180°. The result means that the sign of the vibration displacement is reversed when the magnetic field is reversed. Figure 4.7b shows the H dependence of A_{peak} and ϕ_{peak}. ϕ_{peak} changes by 180° at around $H = 0$ and the amplitude of A_{peak} increases with the application of H. By using these two values, we defined the signed vibration displacement $d_{peak} = A_{peak}\cos\phi_{peak}$ and plotted the H dependence of d_{peak} in Fig. 4.7c (blue dots). d_{peak} increases almost linearly with the increase in H. Comparing the result with the M-H curve of the $Tb_{0.3}Dy_{0.7}Fe_2$ film [inset in Fig. 4.7c], the H dependence of d_{peak} shows similar behavior to the magnetization curve, where M increases with the increase in H before the magnetization saturation. This result suggests that the observed mechanical signal corresponds to the magnetization process of the $Tb_{0.3}Dy_{0.7}Fe_2$ film. As shown in the H dependence of d_{peak} at $I_{AC} = 0, 10, 20, 30, 40, 50$ mA [Fig. 4.7c], the amplitude of d_{peak}

monotonically increases with the increase in I_{AC}, where A_{peak} is proportional to I_{AC} [Fig. 4.7d].

4.3.3 Paramagnetic Metal Dependence

To clarify the origin of the observed mechanical vibration, we checked the influence of the spin current injection on the observed signal by changing the paramagnetic metal layer from Pt to W and Cu. W has a large spin Hall angle similar to that of Pt, but its sign is known to be opposite to that of Pt [10]. Cu shows a small spin Hall angle, and it is possible to remove the contribution of a spin current. The fact means that the paramagnetic metal dependence enables us to verify the influence of the spin Hall effect. Figure 4.8a shows the H dependence of A_{peak} and ϕ_{peak} for the $Tb_{0.3}Dy_{0.7}Fe_2$(100 nm)/W(100 nm) sample (green dots). Compared to the results for the $Tb_{0.3}Dy_{0.7}Fe_2$/Pt sample (blue dots), ϕ_{peak} is reversed by 180°, while A_{peak} increases with H as that for the $Tb_{0.3}Dy_{0.7}Fe_2$/Pt sample. The observed result implies that the sign of the mechanical displacement is opposite to the $Tb_{0.3}Dy_{0.7}Fe_2$/Pt sample. As shown in Fig. 4.8b, this behavior is clearly seen when we plot the H dependence of d_{peak} (green dots), which agrees with the fact that W has a spin Hall angle with the opposite sign to that of Pt. In the case of the $Tb_{0.3}Dy_{0.7}Fe_2$(100 nm)/Cu(100 nm) sample, the amplitude of A_{peak} is strongly suppressed and ϕ_{peak} is almost constant even when the magnetic field is reversed [grey dots in Fig. 4.8a]. The value of d_{peak}, as shown by grey dots in Fig. 4.8b, is small and remains unchanged with the increase in H, consistent with the small spin Hall angle in Cu. These results support that the observed mechanical vibration is attributed to the spin current injection via the spin Hall effect.

4.3.4 Magnetic Field Angle Dependence

Figure 4.9 shows the vibration displacement d_{peak} as a function of the relative angle θ between the a.c. current and the magnetic field. We found that d_{peak} exhibits a clear $\sin\theta$ dependence for the $Tb_{0.3}Dy_{0.7}Fe_2$/Pt sample (blue dots). When Pt is replaced with W (green dots) [Cu (grey dots)], the value of d_{peak} is reversed (suppressed), which satisfies the characteristic of the spin Hall effect.

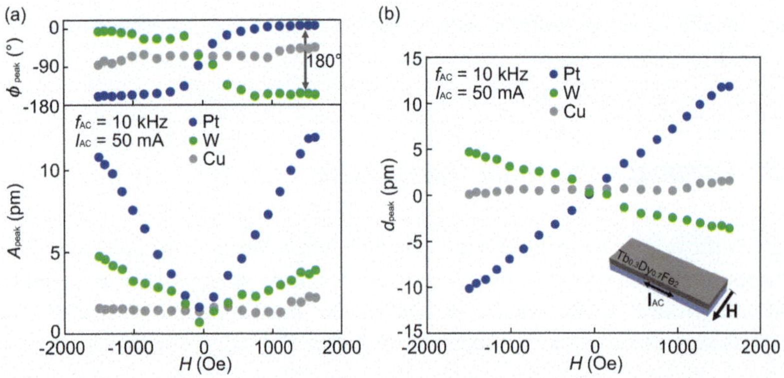

Fig. 4.8 **a** The magnetic field H dependence of A_{peak} and ϕ_{peak} for the $Tb_{0.3}Dy_{0.7}Fe_2$/Pt (blue dots), $Tb_{0.3}Dy_{0.7}Fe_2$/W (green dots), and $Tb_{0.3}Dy_{0.7}Fe_2$/Cu (grey dots) samples. A_{peak} and ϕ_{peak} denote the vibration amplitude and phase at $f = f_{AC}$, respectively. **b** The H dependence of $d_{peak} = A_{peak}\cos\phi_{peak}$ for the $Tb_{0.3}Dy_{0.7}Fe_2$/Pt (blue dots), $Tb_{0.3}Dy_{0.7}Fe_2$/W (green dots), and $Tb_{0.3}Dy_{0.7}Fe_2$/Cu (grey dots) samples. The values of the frequency f_{AC} and the intensity I_{AC} of the applied a.c. current were fixed at 10 kHz and 50 mA, respectively. Adapted from Nat. Commun. **13**, 2440 (2022) [4]; copyright 2022 Author(s), licensed under a Creative Commons Attribution (CC BY) license

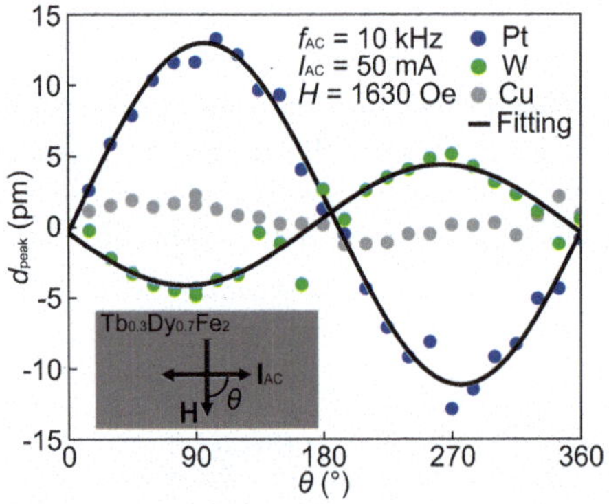

Fig. 4.9 The θ dependence of $d_{peak} = A_{peak}\cos\phi_{peak}$ for the $Tb_{0.3}Dy_{0.7}Fe_2$/Pt (blue dots), $Tb_{0.3}Dy_{0.7}Fe_2$/W (green dots), and $Tb_{0.3}Dy_{0.7}Fe_2$/Cu (grey dots) samples. θ is defined as the relative angle between the magnetic field $\mathbf{H}$ and the a.c. current $\mathbf{I}_{AC}$. The black solid curve is a fitting. The frequency f_{AC} of the a.c. current, the intensity I_{AC} of the a.c. current, and the magnetic field H were set to 10 kHz, 50 mA, and 1630 Oe, respectively. Adapted from Nat. Commun. **13**, 2440 (2022) [4]; copyright 2022 Author(s), licensed under a Creative Commons Attribution (CC BY) license

4.4 Theoretical Calculation of Spin Current Volume Effect

The experimental results suggest that the observed signal is attributable to spin-current-induced mechanical responses in the $Tb_{0.3}Dy_{0.7}Fe_2$ film. In this section, we construct a theoretical model of the magnetostriction in response to the spin current injection, and analyze the obtained experimental results based on the model.

4.4.1 Modulation of Magnetization Angle

We first calculate the equilibrium state of magnetization under spin transfer torque [11, 12]. We start with the LLG equation [Eq. (1.10)] with the spin-transfer torque $\boldsymbol{\tau}_{\rm stt}$:

$$\frac{d\mathbf{M}}{dt} = -\gamma\mathbf{M}\times\mathbf{H}_{\rm ext} + \frac{\alpha}{M_{\rm s}}\mathbf{M}\times\frac{d\mathbf{M}}{dt} + \boldsymbol{\tau}_{\rm stt} \tag{4.1}$$

$$= -\gamma\mathbf{M}\times\mathbf{H}_{\rm ext} + \frac{\alpha}{M_{\rm s}}\mathbf{M}\times\frac{d\mathbf{M}}{dt} + \frac{\gamma j_{\rm s}}{M_{\rm s}^2 d}\mathbf{M}\times(\mathbf{M}\times\hat{\boldsymbol{\sigma}}), \tag{4.2}$$

$$\mathbf{H}_{\rm ext} = (0, 0, H_{\rm ext}), \tag{4.3}$$

$$\hat{\boldsymbol{\sigma}} = (\cos\theta, 0, -\sin\theta). \tag{4.4}$$

Here, $\mathbf{M}$, $\mathbf{H}_{\rm ext}$, $\hat{\boldsymbol{\sigma}}$, $j_{\rm s}$, $M_{\rm s}$, d, and θ are the magnetization of the $Tb_{0.3}Dy_{0.7}Fe_2$ film, the external magnetic field, the spin accumulation at the interface, a spin current, the saturation magnetization of the $Tb_{0.3}Dy_{0.7}Fe_2$ film, the thickness of the $Tb_{0.3}Dy_{0.7}Fe_2$ film, and the relative angle between an electric current $\mathbf{j}_{\rm c}$ and the external field, respectively. Figure 4.10 shows a schematic illustration of the calculation model. The following calculation uses the $x'y'z'$ coordinate system where $z' \parallel \mathbf{H}_{\rm ext}$. While $\mathbf{M} \parallel \mathbf{H}_{\rm ext}$ at $\boldsymbol{\tau}_{\rm stt} = 0$ in equilibrium, the magnetization direction changes at $\boldsymbol{\tau}_{\rm stt} \neq 0$ due to the effective magnetic field by $\boldsymbol{\tau}_{\rm stt}$. The solution $\mathbf{M}^{\rm eq}$ is obtained by substituting $d\mathbf{M}^{\rm eq}/dt = 0$ to Eq. (4.2):

$$0 = -\gamma\mathbf{M}_{\rm eq}\times\mathbf{H}_{\rm ext} + \frac{\gamma j_{\rm s}}{M_{\rm s}^2 d}\mathbf{M}_{\rm eq}\times(\mathbf{M}_{\rm eq}\times\hat{\boldsymbol{\sigma}}) \tag{4.5}$$

When we describe the above equation with the explicit components $\mathbf{M}^{\rm eq} = (M_x^{\rm eq}, M_y^{\rm eq}, M_z^{\rm eq})$, we can obtain

$$0 = -\gamma H_{\rm ext}M_y^{\rm eq} - \frac{\gamma j_{\rm s}}{M_{\rm s}^2 d}(M_z^{\rm eq}M_x^{\rm eq}\sin\theta + (M_y^{\rm eq})^2\cos\theta + (M_z^{\rm eq})^2\cos\theta) \tag{4.6}$$

$$0 = \gamma H_{\rm ext}M_x^{\rm eq} - \frac{\gamma j_{\rm s}}{M_{\rm s}^2 d}(M_y^{\rm eq}M_z^{\rm eq}\sin\theta - M_x^{\rm eq}M_y^{\rm eq}\cos\theta) \tag{4.7}$$

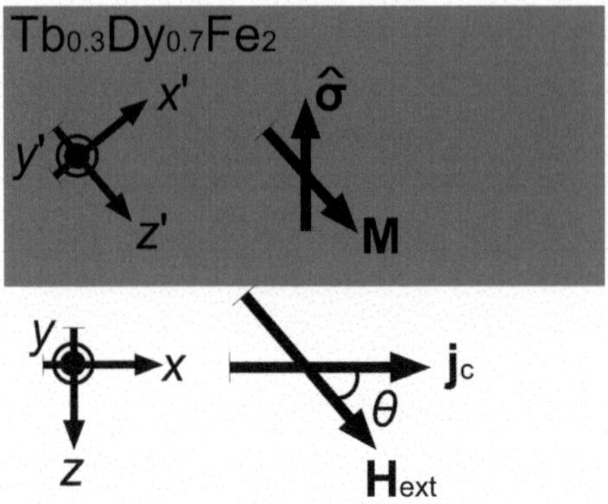

Fig. 4.10 A schematic illustration of the theoretical model. When an electric current $\mathbf{j}_c$ is applied to a paramagnetic metal, the spin accumulation $\hat{\sigma}$ due to the spin Hall effect exerts the spin transfer torque on the magnetization $\mathbf{M}$ of the $Tb_{0.3}Dy_{0.7}Fe_2$ film. θ is the relative angle between the magnetic field $\mathbf{H}_{ext}$ and $\mathbf{j}_c$. We used the $x'y'z'$ coordinate system where $z' \parallel \mathbf{H}_{ext}$

$$0 = \frac{\gamma j_s}{M_s^2 d}((M_x^{eq})^2 \sin\theta + (M_y^{eq})^2 \sin\theta + M_z^{eq} M_x^{eq} \cos\theta) \tag{4.8}$$

The constraint condition $(M_x^{eq})^2 + (M_y^{eq})^2 + (M_z^{eq})^2 = M_s^2$ leads to

$$0 = -\gamma H_{ext} M_y^{eq} - \frac{\gamma j_s}{M_s^2 d}(M_z^{eq} M_x^{eq} \sin\theta + (M_y^{eq})^2 \cos\theta + (M_z^{eq})^2 \cos\theta) \tag{4.9}$$

$$0 = \gamma H_{ext} M_x^{eq} - \frac{\gamma j_s}{M_s^2 d}(M_y^{eq} M_z^{eq} \sin\theta - M_x^{eq} M_y^{eq} \cos\theta) \tag{4.10}$$

$$0 = \frac{\gamma j_s}{M_s^2 d}((M_s)^2 \sin\theta - (M_z^{eq})^2 \sin\theta + M_z^{eq} M_x^{eq} \cos\theta) \tag{4.11}$$

From the above equations we can calculate the leading term of M_i^{eq} as

$$M_x^{eq} \approx -\frac{j_s^2}{H_{ext}^2 d^2 M_s} \sin\theta\cos\theta \tag{4.12}$$

$$M_y^{eq} \approx -\frac{j_s}{H_{ext} d} \cos\theta \tag{4.13}$$

$$M_z^{eq} \approx M_s \tag{4.14}$$

The obtained result means that the x and y components appear due to the spin transfer torque by the spin current injection. In other words, the effective magnetic field by the spin accumulation tilts the magnetization from the external magnetic field direction.

4.4.2 Modulation of Thermal Fluctuations of Magnetization

The spin transfer torque modulates the magnetization fluctuation as well as the magnetization direction. In this subsection, we calculate the spin-current-induced magnetization modulation by considering the magnetization fluctuation.

Stochastic LLG Equation and Linear Response

To take the thermal fluctuation of magnetization into account, we use the stochastic LLG equation [13], which includes the random magnetic field $\mathbf{h}(t) = (h_x(t), h_y(t), h_z(t))$:

$$\frac{d\mathbf{M}}{dt} = -\gamma\mathbf{M} \times (\mathbf{H}_{\text{ext}} + \mathbf{h}(t)) + \frac{\alpha}{M_{\text{s}}}\mathbf{M} \times \frac{d\mathbf{M}}{dt} + \frac{\gamma j_{\text{s}}}{M_{\text{s}}^2 d}\mathbf{M} \times (\mathbf{M} \times \hat{\sigma}) \tag{4.15}$$

$$\mathbf{H}_{\text{ext}} = (0, 0, H_{\text{ext}}) \tag{4.16}$$

$$\hat{\sigma} = (\cos\theta, 0, -\sin\theta) \tag{4.17}$$

Here, $\mathbf{h}(t)$ satisfies

$$\langle h_i(t)h_j(t')\rangle = \frac{2\alpha k_{\text{B}}T}{\gamma M_{\text{s}}V}\delta_{i,j}\delta(t-t'), \tag{4.18}$$

where, k_{B}, T, V are the Boltzmann constant, temperature, and the magnetic coherence volume of the $Tb_{0.3}Dy_{0.7}Fe_2$ film, respectively, and $\langle f \rangle$ denotes the statistical average of f. Equations (4.15) and (4.18) mean that the magnetization $\mathbf{M}$ undergoes the magnetic torque from $\mathbf{h}(t)$ that has no time correlation. As is similar to the previous section, we describe Eq. (4.15) with the magnetization components $\mathbf{M} = (M_x, M_y, M_z)$ as

$$\frac{dM_x}{dt} = -\gamma H_{\text{ext}}M_y - \gamma(M_y h_z(t) - M_z h_y(t)) + \frac{\alpha}{M_{\text{s}}}(M_y\frac{dM_z}{dt} - M_z\frac{dM_y}{dt}) - \frac{\gamma j_{\text{s}}}{M_{\text{s}}^2 d}(M_z M_x\sin\theta + M_y^2\cos\theta + M_z^2\cos\theta) \tag{4.19}$$

$$\frac{dM_y}{dt} = \gamma H_{\text{ext}}M_x - \gamma(M_z h_x(t) - M_x h_z(t)) + \frac{\alpha}{M_{\text{s}}}(M_z\frac{dM_x}{dt} - M_x\frac{dM_z}{dt}) - \frac{\gamma j_{\text{s}}}{M_{\text{s}}^2 d}(M_y M_z\sin\theta - M_x M_y\cos\theta) \tag{4.20}$$

$$\frac{dM_z}{dt} = \gamma(M_x h_y(t) - M_y h_x(t)) + \frac{\alpha}{M_{\text{s}}}(M_x\frac{dM_y}{dt} - M_y\frac{dM_x}{dt}) + \frac{\gamma j_{\text{s}}}{M_{\text{s}}^2 d}(M_x^2\sin\theta + M_y^2\sin\theta + M_z M_x\cos\theta) \tag{4.21}$$

For simplicity, we assume that the magnetization change parallel to the field is negligibly small $dM_z/dt \approx 0$. Furthermore, considering the present situation of $M_z \gg M_x, M_y$, we perform the first-order approximation about M_x, M_y:

$$\frac{dM_x}{dt} \approx -\gamma H_{\text{ext}} M_y + \gamma M_z h_y(t) - \frac{\alpha}{M_s} M_z \frac{dM_y}{dt} - \frac{\gamma j_s}{M_s^2 d}(M_z M_x \sin\theta + M_z^2 \cos\theta) \quad (4.22)$$

$$\frac{dM_y}{dt} \approx \gamma H_{\text{ext}} M_x - \gamma M_z h_x(t) + \frac{\alpha}{M_s} M_z \frac{dM_x}{dt} - \frac{\gamma j_s}{M_s^2 d} M_y M_z \sin\theta \quad (4.23)$$

$$\frac{dM_z}{dt} \approx 0 \quad (4.24)$$

When $\mathbf{h}(t) = 0$, the equilibrium state $dM_i/dt = 0$ is fulfilled after sufficient time, but in the present case of $\mathbf{h}(t) \neq 0$, the magnetization precession components (M_x, M_y) remain temporally changing. To extract this time-dependent term due to the magnetization fluctuation, we define the magnetization components $m_x(t), m_y(t)$ as

$$\mathbf{M}(t) = (M_x^{\text{eq}} + m_x(t), M_y^{\text{eq}} + m_y(t), M_z) \quad (4.25)$$

$$= \left(-\frac{j_s^2}{H_{\text{ext}}^2 d^2 M_s} \sin\theta\cos\theta + m_x(t), -\frac{j_s}{H_{\text{ext}} d}\cos\theta + m_y(t), M_z\right) \quad (4.26)$$

Here, $M_x^{\text{eq}}, M_y^{\text{eq}}$ are the magnetization components in equilibrium when $\mathbf{h}(t) = 0$. By substituting the above equation to the stochastic LLG equation, we can obtain

$$\frac{dm_x}{dt} = -\gamma H_{\text{ext}}\left(-\frac{j_s}{H_{\text{ext}} d}\cos\theta + m_y(t)\right) + \gamma M_z h_y(t) - \frac{\alpha}{M_s} M_z \frac{dm_y}{dt}$$
$$- \frac{\gamma j_s}{M_s^2 d}\left(M_z\left(-\frac{j_s^2}{H_{\text{ext}}^2 d^2 M_s}\sin\theta\cos\theta + m_x(t)\right)\sin\theta + M_z^2\cos\theta\right) \quad (4.27)$$

$$\frac{dm_y}{dt} = \gamma H_{\text{ext}}\left(-\frac{j_s^2}{H_{\text{ext}}^2 d^2 M_s}\sin\theta\cos\theta + m_x(t)\right) - \gamma M_z h_x(t) + \frac{\alpha}{M_s} M_z \frac{dm_x}{dt}$$
$$- \frac{\gamma j_s}{M_s^2 d}\left(-\frac{j_s}{H_{\text{ext}} d}\cos\theta + m_y(t)\right) M_z \sin\theta. \quad (4.28)$$

By assuming that $M_z \approx M_s$ and performing the first order approximation about j_s, the above equations lead to

$$\frac{dm_x}{dt} \approx -\gamma H_{\text{ext}} m_y(t) + \gamma M_z h_y(t) - \frac{\alpha M_z}{M_s}\frac{dm_y}{dt} - \frac{\gamma j_s M_z}{M_s^2 d} m_x(t)\sin\theta, \quad (4.29)$$

$$\frac{dm_y}{dt} \approx \gamma H_{\text{ext}} m_x(t) - \gamma M_z h_x(t) + \frac{\alpha M_z}{M_s}\frac{dm_x}{dt} - \frac{\gamma j_s M_z}{M_s^2 d} m_y(t)\sin\theta. \quad (4.30)$$

These equations give the linear response between $(m_x(t), m_y(t))$ and $(h_x(t), h_y(t))$:

$$\left(\frac{d}{dt} + \frac{\gamma M_z A_{\text{stt}}(\theta)}{M_s}\right) m_x + \left(\gamma H_{\text{ext}} + \frac{\alpha M_z}{M_s}\frac{d}{dt}\right) m_y \approx \gamma M_z h_y(t), \tag{4.31}$$

$$\left(\frac{d}{dt} + \frac{\gamma M_z A_{\text{stt}}(\theta)}{M_s}\right) m_y - \left(\gamma H_{\text{ext}} + \frac{\alpha M_z}{M_s}\frac{d}{dt}\right) m_x \approx -\gamma M_z h_x(t). \tag{4.32}$$

Here we defined

$$A_{\text{stt}}(\theta) = \frac{j_s \sin\theta}{M_s d}. \tag{4.33}$$

Equations (4.31) and (4.32) describe the thermal magnetization precession under the spin transfer torque.

Calculation of Statistic Average of Magnetization

By combining Eqs. (4.31), (4.32) and the linear response theory, we calculate the magnetization modulation by the spin transfer torque in the presence of thermal magnetization fluctuations. The statistic average of $m_i(t)$ is obtained as

$$\langle m_i(t) m_j(0) \rangle = \frac{\alpha k_B T}{\pi \gamma M_s V} \int_{-\infty}^{\infty} \sum_k \chi_{ik}(\omega) \chi^*_{jk}(\omega) e^{-i\omega t} d\omega, \tag{4.34}$$

where $\chi_{ij}(\omega)$ is the susceptibility tensor $\chi(\omega) = [\chi_{ij}(\omega)]$ that relates the linear response

$$\begin{pmatrix} \tilde{m}_x \\ \tilde{m}_y \end{pmatrix} = \chi(\omega) \begin{pmatrix} \tilde{h}_x \\ \tilde{h}_y \end{pmatrix} \tag{4.35}$$

Here,

$$\tilde{m}_i(\omega) = \int_{-\infty}^{\infty} m_i(t) e^{i\omega t} dt, \tag{4.36}$$

$$\tilde{h}_i(\omega) = \int_{-\infty}^{\infty} h_i(t) e^{i\omega t} dt \tag{4.37}$$

are the Fourier transform of $m_i(t)$ and $h_i(t)$. We first describe the linear response relation (4.31) and (4.32) in the frequency space by Fourier transformation:

$$(-i\omega + \frac{\gamma M_z A_{\text{stt}}(\theta)}{M_s})\tilde{m}_x(\omega) + (\gamma H_{\text{ext}} - i\omega \frac{\alpha M_z}{M_s})\tilde{m}_y(\omega) = \gamma M_z \tilde{h}_y(\omega) \tag{4.38}$$

$$(-i\omega + \frac{\gamma M_z A_{\text{stt}}(\theta)}{M_s})\tilde{m}_y(\omega) - (\gamma H_{\text{ext}} - i\omega \frac{\alpha M_z}{M_s})\tilde{m}_x(\omega) = -\gamma M_z \tilde{h}_x(\omega) \tag{4.39}$$

In the matrix form, we can obtain

$$\begin{pmatrix} -\gamma H_{\text{ext}} + i\omega \frac{\alpha M_z}{M_s} & -i\omega + \frac{\gamma M_z A_{\text{stt}}(\theta)}{M_s} \\ i\omega - \frac{\gamma M_z A_{\text{stt}}(\theta)}{M_s} & -\gamma H_{\text{ext}} + i\omega \frac{\alpha M_z}{M_s} \end{pmatrix} \begin{pmatrix} \tilde{m}_x \\ \tilde{m}_y \end{pmatrix} = -\gamma M_z \begin{pmatrix} \tilde{h}_x \\ \tilde{h}_y \end{pmatrix}. \tag{4.40}$$

From the above equation, $\tilde{m}_i$ is written as

$$\begin{pmatrix} \tilde{m}_x \\ \tilde{m}_y \end{pmatrix} = -\gamma M_z \begin{pmatrix} -\gamma H_{\text{ext}} + i\omega \frac{\alpha M_z}{M_s} & -i\omega + \frac{\gamma M_z A_{\text{stt}}(\theta)}{M_s} \\ i\omega - \frac{\gamma M_z A_{\text{stt}}(\theta)}{M_s} & -\gamma H_{\text{ext}} + i\omega \frac{\alpha M_z}{M_s} \end{pmatrix}^{-1} \begin{pmatrix} \tilde{h}_x \\ \tilde{h}_y \end{pmatrix} \tag{4.41}$$

$$= \frac{\gamma M_z}{(\gamma H_{\text{ext}} - i\omega \frac{\alpha M_z}{M_s})^2 + (i\omega - \frac{\gamma M_z A_{\text{stt}}(\theta)}{M_s})^2} \times$$

$$\begin{pmatrix} \gamma H_{\text{ext}} - i\omega \frac{\alpha M_z}{M_s} & -i\omega + \frac{\gamma M_z A_{\text{stt}}(\theta)}{M_s} \\ i\omega - \frac{\gamma M_z A_{\text{stt}}(\theta)}{M_s} & \gamma H_{\text{ext}} - i\omega \frac{\alpha M_z}{M_s} \end{pmatrix} \begin{pmatrix} \tilde{h}_x \\ \tilde{h}_y \end{pmatrix}, \tag{4.42}$$

and the explicit form of $\chi(\omega)$ is

$$\chi(\omega) = \frac{\gamma M_z}{(\gamma H_{\text{ext}} - i\omega \frac{\alpha M_z}{M_s})^2 + (i\omega - \frac{\gamma M_z A_{\text{stt}}(\theta)}{M_s})^2} \times \begin{pmatrix} \gamma H_{\text{ext}} - i\omega \frac{\alpha M_z}{M_s} & -i\omega + \frac{\gamma M_z A_{\text{stt}}(\theta)}{M_s} \\ i\omega - \frac{\gamma M_z A_{\text{stt}}(\theta)}{M_s} & \gamma H_{\text{ext}} - i\omega \frac{\alpha M_z}{M_s} \end{pmatrix} \tag{4.43}$$

By substituting $\chi(\omega)$ to Eq. (4.34), we calculate the statistical average of the magnetization thermal fluctuation components as

$$\langle m_x^2 \rangle = \langle m_x(0) m_x(0) \rangle \tag{4.44}$$

$$= \frac{\alpha k_B T}{\pi \gamma M_s V} \int_{-\infty}^{\infty} (\chi_{xx}\chi_{xx}^* + \chi_{xy}\chi_{xy}^*) d\omega \tag{4.45}$$

When we neglect the second or higher order of α and $A_{\text{stt}}(\theta)$ under the assumption of α, $A_{\text{stt}}(\theta)/H_{\text{ext}} \ll 1$, the above equation is simplified to

$$\langle m_x^2 \rangle \approx (M_z)^2 \frac{2\alpha k_B T}{M_s H_{\text{ext}} V} \int_{-\infty}^{\infty} \frac{d\omega}{2\pi} \frac{\omega_H(\omega^2 + \omega_H^2)}{((\omega_H + \omega)^2 + \Gamma_{\text{stt}}^2(\theta))((\omega_H - \omega)^2 + \Gamma_{\text{stt}}^2(\theta))} \tag{4.46}$$

$$= 2(M_z)^2 \frac{2\alpha k_B T}{M_s H_{\text{ext}} V} \int_0^{\infty} \frac{d\omega}{2\pi} \frac{\omega_H(\omega^2 + \omega_H^2)}{((\omega_H + \omega)^2 + \Gamma_{\text{stt}}^2(\theta))((\omega_H - \omega)^2 + \Gamma_{\text{stt}}^2(\theta))}, \tag{4.47}$$

where we defined

$$\omega_{\mathrm{H}} = \gamma H_{\mathrm{ext}} \tag{4.48}$$

$$\Gamma_{\mathrm{stt}}(\theta) = \frac{\omega_{\mathrm{H}} M_z}{M_{\mathrm{s}}}\left(\alpha + \frac{A_{\mathrm{stt}}(\theta)}{H_{\mathrm{ext}}}\right) \tag{4.49}$$

and took integral interval in $\omega > 0$ considering that the integrant is even function about ω. Here, the part of integrant $\frac{1}{(\omega_{\mathrm{H}}-\omega)^2+\Gamma_{\mathrm{stt}}^2(\theta)}$ takes large values at $\omega = \omega_{\mathrm{H}}$, and we approximately replace ω with ω_{H} in other parts of the integrant as

$$\langle m_x^2 \rangle \approx 2(M_z)^2 \frac{2\alpha k_{\mathrm{B}} T}{M_{\mathrm{s}} H_{\mathrm{ext}} V} \frac{2\omega_{\mathrm{H}}^3}{4\omega_{\mathrm{H}}^2 + \Gamma_{\mathrm{stt}}^2(\theta)} \int_0^{\infty} \frac{d\omega}{2\pi} \frac{1}{(\omega_{\mathrm{H}} - \omega)^2 + \Gamma_{\mathrm{stt}}^2(\theta)} \tag{4.50}$$

$$\approx 2(M_z)^2 \frac{2\alpha k_{\mathrm{B}} T}{M_{\mathrm{s}} H_{\mathrm{ext}} V} \frac{\omega_{\mathrm{H}}}{2} \left(\frac{1}{2\Gamma_{\mathrm{stt}}(\theta)}\right) \tag{4.51}$$

$$\approx M_{\mathrm{s}}^2 \frac{\alpha k_{\mathrm{B}} T \omega_{\mathrm{H}}}{M_{\mathrm{s}} H_{\mathrm{ext}} V \Gamma_{\mathrm{stt}}(\theta)}. \tag{4.52}$$

In the above calculation, we used $\frac{2\omega_{\mathrm{H}}^3}{4\omega_{\mathrm{H}}^2+\Gamma_{\mathrm{stt}}^2(\theta)} \approx \frac{2\omega_{\mathrm{H}}^3}{4\omega_{\mathrm{H}}^2}$. The result means that a finite magnetization thermal precession component $\propto T$ is modulated by the spin current injection in the presence of the spin transfer torque term $\Gamma_{\mathrm{stt}}(\theta)$. Note that $\langle m_y^2 \rangle = \langle m_x^2 \rangle$. From these results we can calculate the average of the z component of the magnetization as

$$\langle M_z \rangle^2 \approx \langle M_z^2 \rangle \tag{4.53}$$

$$= \langle M_{\mathrm{s}}^2 - (M_x^{\mathrm{eq}} + m_x)^2 - (M_y^{\mathrm{eq}} + m_y)^2 \rangle \tag{4.54}$$

$$= M_{\mathrm{s}}^2 - (M_x^{\mathrm{eq}})^2 - (M_y^{\mathrm{eq}})^2 - \langle m_x^2 + m_y^2 \rangle \tag{4.55}$$

$$= M_{\mathrm{s}}^2 - \left(-\frac{j_{\mathrm{s}}^2}{H_{\mathrm{ext}}^2 d^2 M_{\mathrm{s}}} \sin\theta\cos\theta\right)^2 - \left(-\frac{j_{\mathrm{s}}}{H_{\mathrm{ext}} d}\cos\theta\right)^2 - 2M_{\mathrm{s}}^2 \frac{\alpha k_{\mathrm{B}} T \omega_{\mathrm{H}}}{M_{\mathrm{s}} H_{\mathrm{ext}} V \Gamma_{\mathrm{stt}}(\theta)} \tag{4.56}$$

Here we assumed that the deviation of M_z is small and $\langle M_z^2 \rangle = \langle M_z \rangle^2 + \langle (M_z - \langle M_z \rangle)^2 \rangle \approx \langle M_z \rangle^2$.

By solving the equation, we can obtain

$$\langle M_z \rangle \approx M_s - M_s \frac{\alpha k_B T}{M_s H_{\text{ext}} V(\alpha + \frac{A_{\text{stt}}(\theta)}{H_{\text{ext}}})} \tag{4.57}$$

$$= M_s \left(1 - \frac{\alpha k_B T}{M_s H_{\text{ext}} V(\alpha + \frac{A_{\text{stt}}(\theta)}{H_{\text{ext}}})}\right) \tag{4.58}$$

$$\approx M_s \left(1 - \frac{k_B T(1 - \frac{A_{\text{stt}}(\theta)}{\alpha H_{\text{ext}}})}{M_s H_{\text{ext}} V}\right) \tag{4.59}$$

$$= M_s \left(1 - \frac{k_B T}{M_s H_{\text{ext}} V}\right) + \frac{k_B T A_{\text{stt}}(\theta)}{\alpha H_{\text{ext}}^2 V} \tag{4.60}$$

$$= M_s(T) + \Delta M_z^{\text{stt}}. \tag{4.61}$$

Here we defined the saturation magnetization $M_s(T)$ at temperature T and the magnetization modulation ΔM_z^{stt} due to the spin transfer torque as

$$M_s(T) = M_s \left(1 - \frac{k_B T}{M_s H_{\text{ext}} V}\right), \tag{4.62}$$

$$\Delta M_z^{\text{stt}} = \frac{k_B T A_{\text{stt}}(\theta)}{\alpha H_{\text{ext}}^2 V} = \frac{j_s k_B T}{\alpha H_{\text{ext}}^2 M_s V d} \sin\theta. \tag{4.63}$$

The obtained result means that the magnetization amplitude changes by ΔM_z^{stt} due to the magnetization fluctuation modulation via the spin current injection.

4.4.3 Spin Current Volume Effect by Magnetization Fluctuation Modulation

The magnetization modulation by the spin transfer torque triggers magnetostriction via the spin-lattice coupling. In this section, we calculate the magnetostriction caused by the magnetization modulation based on the free energy. Based on the fact that $Tb_{0.3}Dy_{0.7}Fe_2$ is a ferromagnet that exhibits a morphotropic phase transition, we define the free energy of the system as follows [14, 15]:

$$F = F_{\text{m}} + F_{\text{el}} + F_{\text{m-el}} \tag{4.64}$$

$$F_{\text{m}} = F_{\text{m}}(m_i) \tag{4.65}$$

$$F_{\text{el}} = \frac{c_{11}^0}{2}(e_1^2 + e_2^2 + e_3^2) + c_{12}^0(e_1 e_2 + e_2 e_3 + e_3 e_1) + \frac{c_{44}^0}{2}(e_4^2 + e_5^2 + e_6^2) \tag{4.66}$$

$$F_{\text{m-el}} = Q_{11}^0(e_1 m_1^2 + e_2 m_2^2 + e_3 m_3^2) + Q_{12}^0(e_1(m_2^2 + m_3^2) + e_2(m_3^2 + m_1^2) + e_3(m_1^2 + m_2^2)) + Q_{44}^0(e_4 m_2 m_3 + e_5 m_3 m_1 + e_6 m_1 m_2), \tag{4.67}$$

where m_i and e_i are the i components of the magnetization and strain, respectively. Here we used the Voigt notation. Each term of F_m, F_{el}, and F_{p-e} corresponds to the magnetic energy, the elastic energy, and the magnetoelastic energy, respectively. By imposing the local minimum condition about e_i, $\partial F/\partial e_i = 0$, we can obtain the strain tensor as a function of the magnetization m_i as

$$e_{ii} = -\frac{c_{11}^0 Q_{11}^0 + c_{12}^0 Q_{11}^0 - 2c_{12}^0 Q_{12}^0}{(c_{11}^0 - c_{12}^0)(c_{11}^0 + 2c_{12}^0)} m_i^2 - \frac{c_{11}^0 Q_{12}^0 - c_{12}^0 Q_{11}^0}{(c_{11}^0 - c_{12}^0)(c_{11}^0 + 2c_{12}^0)} (m_{i+1}^2 + m_{i+2}^2), \tag{4.68}$$

$$e_{ij} = -\frac{Q_{44}^0}{c_{44}^0} m_i m_j \quad (i \neq j)\,. \tag{4.69}$$

By substituting the magnetization

$$\langle M_x \rangle = M_x^{\mathrm{eq}} \tag{4.70}$$

$$\langle M_y \rangle = M_y^{\mathrm{eq}} \tag{4.71}$$

$$\langle M_z \rangle = M_s(T) + \Delta M_z^{\mathrm{stt}} \tag{4.72}$$

to Eqs. (4.68) and (4.69), we can formulate the magnetostriction under the spin transfer torque. To compare the experimental result with the calculation, we obtain the out-of-plane displacement component $d_{yy} + d_{yz}$ due to the magnetostriction e_{yy} and e_{yz} in the $Tb_{0.3}Dy_{0.7}Fe_2$ film as

$$\begin{aligned} d_{yy} + d_{yz} \approx\; & d \frac{c_{12}^0 Q_{11}^0 - c_{11}^0 Q_{12}^0}{(c_{11}^0 - c_{12}^0)(c_{11}^0 + 2c_{12}^0)} M_s(T)^2 \\ & + \frac{c_{12}^0 Q_{11}^0 - c_{11}^0 Q_{12}^0}{(c_{11}^0 - c_{12}^0)(c_{11}^0 + 2c_{12}^0)} \frac{2M_s(T) k_B T}{\alpha H_{\mathrm{ext}}^2 M_s V} j_s \sin\theta + \frac{Q_{44}^0 w M_s(T)}{c_{44}^0 H_{\mathrm{ext}} d} j_s \cos\theta \end{aligned} \tag{4.73}$$

Here, we performed the first-order approximation about j_s. w is a prefactor on the size of the $Tb_{0.3}Dy_{0.7}Fe_2$ film. The second (third) term corresponds the magnetostriction due to the magnetization fluctuation modulation $\propto \sin\theta$ (magnetization angle modulation $\propto \cos\theta$) by the spin transfer torque.

4.5 Origin of Observed Mechanical Signal

The observed mechanical vibration peak in the experiments corresponds to the first order with respect to j_s in Eq. (4.73):

$$d_{1\mathrm{st}} = d_{\sin} + d_{\cos} \tag{4.74}$$

$$= \frac{c_{12}^0 Q_{11}^0 - c_{11}^0 Q_{12}^0}{(c_{11}^0 - c_{12}^0)(c_{11}^0 + 2c_{12}^0)} \frac{2M_s(T) k_B T}{\alpha H_{\mathrm{ext}}^2 M_s V} j_s \sin\theta + \frac{Q_{44}^0 w M_s(T)}{c_{44}^0 H_{\mathrm{ext}} d} j_s \cos\theta \tag{4.75}$$

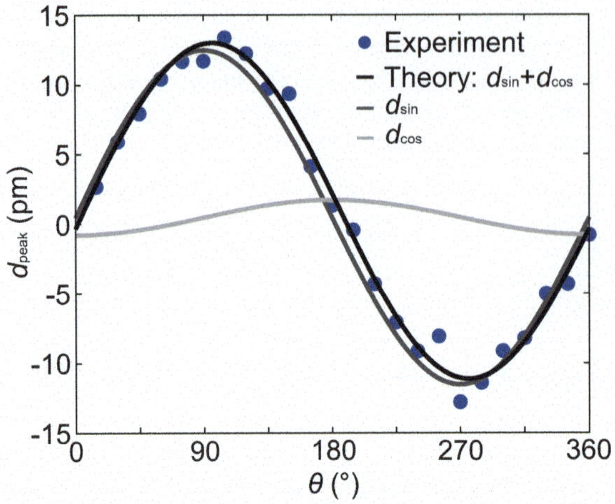

Fig. 4.11 Theoretical fitting of the θ dependence of the mechanical displacement $d_{peak} = A_{peak} \cos\phi_{peak}$ for the $Tb_{0.3}Dy_{0.7}Fe_2$/Pt sample. θ denotes the relative angle between the magnetic field and the a.c. current. The blue dots are the experimental results and the black solid curve is a fitting by the theoretical formula $d_{sin} + d_{cos}$. The first term $d_{sin} \propto \sin\theta$ and the second term $d_{cos} \propto \cos\theta$ in the fitting are shown by grey curves

By comparing the result with the angle θ dependence in the experiment, we found that the first term d_{sin}, which is proportional to $\sin\theta$, is dominant in the present system. Figure 4.11 shows the fitting curves of the experimental result to the theoretical formula (4.75). From the fitting, we confirmed that the ratio of the peak amplitudes of the $\sin\theta$ and $\cos\theta$ components is 1:0.096 in the experiment. These results support our interpretation that the observed mechanical vibration originates from the volume change via the spin-current induced magnetization fluctuation modulation in the $Tb_{0.3}Dy_{0.7}Fe_2$ film.

4.6 Summary

In this study, we demonstrated the spin-current-induced volume modulation effect and clarified its mechanism. By injecting a spin current to a giant magnetostrictive material $Tb_{0.3}Dy_{0.7}Fe_2$, we found that the $Tb_{0.3}Dy_{0.7}Fe_2$ film exhibits the mechanical vibration in response to the spin current injection. Our theoretical analysis revealed that the observed mechanical vibration originates from the magnetization fluctuation modulation due to the spin current injection.

References

1. Physics of ferromagnetism. In: Chikazumi S (ed). Oxford Science Publications, Oxford (2009)
2. Shiga M (1981) J Phys Soc Jpn 50:2573
3. Ando K, Takahashi S, Harii K, Sasage K, Ieda J, Maekawa S, Saitoh E (2008) Phys Rev Lett 101:036601
4. Arisawa H, Shim H, Daimon S, Kikkawa T, Oikawa Y, Takahashi S, Ono T, Saitoh E (2022) Nat Commun 13:2440
5. Materials Ferromagnetic (1980). In: Wohlfarth EP. North-Holland Publishing Company
6. Bergstrom R Jr, Wuttig M, Cullen J, Zavalij P, Briber R, Dennis C, Garlea VO, Laver M (2013) Phys Rev Lett 111:017203
7. Shim H, Sakamoto K, Inomata N, Toda M, Toan NV, Ono T (2020) Micromachines 11:523
8. Clark AE, Abbundi R, Gillmor WR (1978) IEEE Trans Magn MAG-14, 54
9. Sandlund L, Fahlander M, Cedell T, Clark AE, Restorff JB, Wun-Fogle M (1994) J Appl Phys 75:5656
10. Hoffmann A (2013) IEEE Trans Magn 49:5172–5193
11. Slonczewski JC (1996) J Magn Magn Mater 159:L1
12. Berger L (1996) Phys Rev B 54:9353
13. Xiao J, Bauer GEW, Uchida K, Saitoh E, Maekawa S (2010) Phys Rev B 81:214418
14. Ishibashi Y, Iwata M (1997) Jpn J Appl Phys 36:254
15. Ishibashi Y, Iwata M (1999) Jpn J Appl Phys 38:800

Chapter 5
Conclusion

We demonstrated spin-mechanical interaction phenomena by means of a laser Doppler vibrometry, and clarified its mechanism by combining experiments and theoretical analysis.

The first study 'Magnetomechanical coupling in a magnetic micro mechanical oscillator' is summarized in the following [1]:

1. By using the focused ion beam method, we established a microfabrication method for a micro mechanical bridge structure made of a ferrimagnetic insulator $Y_3Fe_5O_{12}$ (YIG).
2. We revealed that the mechanical resonance frequency in the YIG micro bridge changes according to the magnetization process in the bridge. Based on the free energy theory including the magnetoelastic energy, we demonstrated that the observed resonance-frequency change originates from the elastic modulus modulation associated with the magnetization rotation in the YIG bridge.

The second study 'A spin current volume effect in a spin-lattice coupled system' is summarized as follows [2]:

1. We demonstrated the spin-current-induced mechanical response in a giant magnetostrictive material $Tb_{0.3}Dy_{0.7}Fe_2$. By injecting a spin current into a $Tb_{0.3}Dy_{0.7}Fe_2$ film by using the spin Hall effect in paramagnetic films, we found that the $Tb_{0.3}Dy_{0.7}Fe_2$ film exhibits the mechanical vibration that satisfies the characteristic of the spin Hall effect.
2. By comparing experimental results with theoretical analysis, we verified that the observed mechanical vibration in the $Tb_{0.3}Dy_{0.7}Fe_2$ film originates from the volume modulation of the film due to the spin current injection. Our magnetoelastic theoretical modeling revealed that the volume effect is caused by the spin-current-induced magnetization fluctuation modulation.

H. Arisawa, *A Study on Spin Mechanics Phenomena in Spin-Lattice Coupled Systems*, Springer Theses, https://doi.org/10.1007/978-981-97-8642-8_5

These findings accelerate the exploration of cross-correlation phenomena between spintronics and micromechanics and provide insight into magnetomechanical properties associated with a spin current.

References

1. Arisawa H, Daimon S, Oikawa Y, Seo Y-J, Harii K, Oyanagi K, Saitoh E (2019) Appl Phys Lett 114:122402
2. Arisawa H, Shim H, Daimon S, Kikkawa T, Oikawa Y, Takahashi S, Ono T, Saitoh E (2022) Nat Commun 13:2440

The manufacturer's authorised representative in the EU is Springer Nature Customer Service Centre GmbH, Europaplatz 3, 69115 Heidelberg, Germany. If you have any concerns regarding our products, please contact ProductSafety@springernature.com

Printed and bound by CPI Group (UK) Ltd, Croydon, CR0 4YY
01/10/2026
02225980-0003